for Dick Leather —
skipper & friend —
perhaps something to argue about on the dog w

Jerry

Human Transactions

Human Transactions

The Emergence of Meaning in Time

Gary H. Stahl

Temple University Press
Philadelphia

Temple University Press, Philadelphia 19122

Published 1995
Printed in the United States of America

♾ The paper used in this book meets the requirements of the American National Standard for Information Sciences—Permanence of Paper for Printed Library Materials, ANSI Z39.48-1984.

Text design by Dariel Mayer

Library of Congress Cataloging-in-Publication Data

Stahl, Gary.
Human Transactions : the emergence of meaning in time / Gary H. Stahl.
p. cm.
Includes bibliographical references and index.
ISBN 1-56639-287-X (alk. paper)
1. Ethics, Evolutionary. 2. Interpersonal relations—Moral and ethical aspects. 3. Free will and determinism. I. Title.
BJ1311.S76 1995
171'.7—dc20 94-21869

Contents

Acknowledgments

I had the great good fortune to study with two philosophers whose work I deal with, however inadequately, in this book: John W. Miller, professor at Williams College, and Albert Hofstadter, professor at Columbia University. I thank them, too late, for their wisdom.

Dr. John Dickerson, formerly of Trinity College, currently CEO of Dickerson Distributors, brought his reflective intelligence and clarity to bear on each draft of each chapter. The book is stronger and cleaner for his efforts, and richer for his friendship.

I would not have been aware of the expansive relevance of biological and psychological matters without many instructive discussions with two particular friends and colleagues from the department of psychology at the University of Colorado: Professor Louise Silvern (Clinical), whose work is the very model of philosophical reflection, read the manuscript with challenging care; Professor Eugene Gollin (Developmental) and I have fought the good fight over reductionism, mind-body, rats and rationalism, for many years. His wealth of knowledge and methodological precision are a mark to shoot for.

My colleagues in philosophy during the last thirty-three years at the University of Colorado are both too numerous to mention and too important to risk leaving out, so I thank them collectively, but with no less warmth and appreciation, for their friendship and their arguments.

The final—and anonymous—prepublication reviewer was perceptive, tough-minded, and enormously helpful in making the argument stronger and more responsible.

Obviously, none of the above can be blamed for any misinterpretations or confusions with which I may have distorted their sound advice.

I wish to thank the Council on Research and Creative Work at the University of Colorado for a Faculty Fellowship, and the university for a sabbatical leave; this help permitted me to put full effort into the book.

Most fundamentally, I thank my wife, Mitch, and my sons, Bill and Harry, for bringing into my life the kind of rich meaning I have tried to make sense of in this book.

Portions of chapter 1 first appeared as "On Not Reducing Agents to Organisms," in *Journal for the Theory of Social Behavior* 12, no. 3 (1982): 305–16. Reprinted by permission.

Portions of chapter 2 first appeared as "Bioethics: A Redundant Concept," in *Ethics: Foundations, Problems, and Applications,* ed. E. Morscher, Schriftenreihe der Wittgenstein-Gesellschaft 7. Copyright 1981 Holder-Pichler-Tempsky, Vienna, Austria. Reprinted by permission.

Portions of chapter 3 first appeared as "The Completion of the Past," in *The Philosopher in the Community: Essays in Memory of Bertram Morris,* ed. Berel Lang, William Sacksteder, and Gary Stahl (Lanham, Md.: University Press of America, 1984), 168–77. Reprinted by permission.

Portions of chapter 4 first appeared as "Making the Moral World," in *The Philosophy of John William Miller,* ed. Joseph Fell, *Bucknell Review* 34, no. 1 (1990): 111–22. Reprinted by permission.

Portions of chapter 6 first appeared in Eugene Gollin, Gary Stahl, and Elyse Morgan, "On the Uses of the Concept of Normality in Developmental Biology and Psychology," in *Advances in Child Development and Behavior,* vol. 21, ed. Hayne W. Reese (1989): 49–71. Reprinted by permission. Copyright 1989 by Academic Press.

Portions of chapter 8 first appeared as "Art and the Emergence of Self: Reflections on Some Central Themes in the Work of Albert Hofstadter," in *International Philosophical Quarterly* 15, no. 3 (1975): 333–51. Reprinted by permission.

Human Transactions

Re question 1

The human species can reckon with its own problems and arrive at workable relief (justice) from circumstances. Does that necessitate belief in any formulation prior to doing justice?

Re 2

Why does the absence of such norms connote arbitrary/non compelling? Choices can be logical and/or epigenetic.

Re 3

Meaning at the individual level is certain. That the ship is rudderless and without power does not suggest one cannot swab the deck with effect.

Introduction

The Three Original Questions

This book responds to three questions:

1. Is the belief that people can make responsible and autonomous choices about their lives compatible with what we know about the evolutionary and developmental processes out of which people emerge?
2. How can the choices made within biological and cultural history be nonarbitrary and compelling if there are no ahistoric norms against which they can be judged?
3. What sort of meaning can individual lives and choices have if the history within which they come to be is simply an accidental episode in an indifferent universe?

The meanings of these questions will evolve as we work through them, but we can begin by characterizing them as the problems set by organismic restraints, by the absence of norms, and by a lack of purpose in the universe. But however we take them, they are so complex that it is more appropriate to talk about "responding" to them than "answering" them, even in part. All make nonsense of the boundaries between the disciplines, even of the much-touted wall between science and humanities. But they do point toward unavoidable areas of human concern; and within the focus given by these perennial concerns we can put together questions concrete and contextual enough to generate responses that are, in the appropriate sense, rough answers.

In fact (or so I will argue), if we ask the right questions we find that we already know an immense amount of what is needed to make sense of our lives as agents creating history in an indifferent universe. Unfortunately, we keep asking misleading questions, so even our partial stories are couched in different languages, kept in different libraries.

Thus we have a solid scientific grip both on the basic evolutionary processes through which distinctly human animals emerge as a species, and on the developmental processes through which individual members of this species grow from infants to become persons. We also have thousands of years of deep and sometimes dark reflection on the challenges the species faces if it is to flourish, and which face individuals if they are to make sense of their brief lives. Unfortunately (disastrously), we do not see this scientific knowledge of our emergence and the humanistic reflection on its meaning as perspectives on the same process, but as methodologically incompatible explorations of utterly disparate subject matters. I could parody nineteenth-century disputations by describing this as the conflict between objective, quantitative Science which investigates Matter, and subjective qualitative Art, which reflects the vagaries of Mind. I wish this were more of a parody and more behind us than it is.

True enough, the old dichotomies between "hard science" (reporting the facts) and "soft art" (giving the human meaning of the facts) are everywhere dissolving within the various disciplines; but our everyday ways of thinking about the world go on under the old dispensation. Even the professionals, who in their own fields dismiss all the dated dualisms as anachronisms, join the rest of us (just who "we" are will have to be sorted out) in the marketplace, contrasting the public objectivity of science with the personal idiosyncrasy of values, taking the difference as one that goes without thinking. Nor are the universities of much help here, in spite of all their colleges of arts *and* sciences, where only the freshmen take the "and" to be conjunctive.

Humanistic reflection is not always seen as idiosyncratic (and probably dispensable): there is also the powerful and honored view that values are separate from—though probably threatened by—the processes described by science. Such values have a "higher" destiny and source in some absolute realm untouched by mortality and change; both God's mind and Marx's ahistoric script for history give a platform apart from evolution and development from which "objective" judgments can be launched back into time. Unfortunately, such immunity from the vagaries of time rules out the possibility that experiences within time can enrich or revise these supposed grounds of judgment: God's mind does not mature under the pummelings of experience, nor does history change its dialectical destiny aimed at the withering away of the state. This sort of theoretic reassurance disfranchises intelligence, and reads all scientific and responsible evidence out of court.

The flight into the timeless would not seem so attractive, nor the surrender to subjectivity so threatening, if the worlds of science did not appear so alien to all that is distinctly human. From the conventional view of science, the response to the three questions is bleak indeed, and the scientific story of our emergence appears to make nonsense of the traditional notions of autonomy and responsibility, picturing us as territorial apes with programmed software for brains. What can "meaning" be for such creatures but jangling the ganglia to satisfy whatever atomistic urges come with the body or get scheduled in through the accidents of conditioning? What criteria for action can there be but a cost/benefit guess at which way of living will give the most psychic M&Ms?

This view of things is (literally) terrifying enough to send us scurrying back into the timeless. But unnecessarily so, for a more adequate reading of what the sciences say about evolution and development shows that time and change are not threats to our meaning and responsibility, but the condition of our having them. Indeed, the process of becoming a distinctly human animal consists in good part precisely in reflecting on what it means to emerge and become human in a niche where we are vulnerable both to nature and to our own demented wills. Hence with us, as with any finite and embodied self-consciousness, meaning is an achievement, not a given. We can borrow immortality-conferring scripts (Becker 1973) from God or history, but they are never ours, authenticated and revised through our experience, responsive to our deeds.

There is a sense in which every organism's biological animality is a given, simply that which has evolved in a niche and gotten written into its DNA. But a coherent human story that can give our lives a livable unity in the face of inevitable death is not a given that comes with the territory, but a chancy and ever-revisable projection. But here is the old problem back again: such meanings as people cob together one day at time, in time, have been rejected precisely because they are time-bound, vulnerable, subject to change. To generation after generation of plain folk and thinkers alike, it has seemed self-evident that human choices are too tricky and full of consequence to be made on merely human grounds, or in terms of merely human ends: the "merely" human always stinks of mortality and the inevitable failure of all human projects in animal death. And it is with just this animality and this death that the scientific story threatens us. So it is at least partly to escape death, or at least the meaninglessness of a simply animal death, that we are tempted to go beyond science into some timeless tale of essence and eternality. Most of us have been able to live in our space and time within history only by conjuring up some greater context—immune to space and time—within which our history falls, and of which history is the shadowy illustration.

The sticking point is *perspective,* for in all the descriptions science gives of our rise and destined fall, we are stuck in some local mess, struggling to survive and come to be not vis-à-vis the Universe-as-a-Whole, but in some particular niche and cranny. To be thus placed and embodied is to be caught in perspective, to see with *these* eyes from *this* point of view. Given this, what we report of the world must fall short of Truth; the mere images we grasp must fail of Beauty; and the Good for which we strive must remain a dream. To be caught up in perspective is to be blocked forever from the Real.

These consequences do indeed follow: to take ourselves seriously as the animals evolving and developing in the hurly-burly of time *is* to be cast out of the Garden of unearthly delights, and to lose forever the hope that all our strivings will make some ultimate difference, and our deaths none. It *does* mean that we are caught up in a series of perspectives whose revision can never come to a final conclusion, only to an end in time. But what it does not mean is that any perspective is as good as any other, or that the process of revision cannot be critical and responsible. Nor does it mean that the lives we lead lack meaning in the only senses that are relevant and defendable, be the universe as indifferent as it may. The task is to show how this inevitable and inescapable immersion in local times and spaces does not threaten the possibility of a sustainable human life. In terms of our current cultural conflicts, I stand the conservative argument on its head, and argue that values can be authoritative and reliable *only if* they do indeed depend on human actions, and not on fixed and immutable standards.

Mostly, we have not been willing to risk this enterprise of finding meaning in our own biographies and histories. It has seemed easier to accept the supposed verdict of science and reduce the humanistic world to whim. Or we have been willing to accept science as the arbiter of the merely materialistic, but kept value warm and safe in a realm untouchable by the worms of time. What we have not done is allow the scientific ways of thinking to permeate our thinking about value, or to build the implications of evolution and development into our sense of who we are. What it would mean to do so is the theme of this book.

This emphasis on science obviously affects the ways in which the original three questions are phrased, and I will have to deal directly (in chapter 5) with the question of whether illegitimate and question-begging assumptions are involved here. I believe they are not, and that the questions focus on issues that—whatever the phraseology—are fundamental to the human enterprise. In this sense, the argument should be of interest even to those with a theological point of view, since the issue of how one can be free and responsible, yet under the law, is quite inescapable, no matter how divine the law. My argument stakes its claim to be taken seriously on whatever power it has to make sense of these unavoidable issues, both by clearing away barriers to the use of science, and by defining the areas of inquiry within which critical intelligence can sustain our attempts to make meaning within history.

The basic terms of the argument are those of a nonreductive naturalism that sees process and time as fundamental. Although human beings are organisms embodied and embedded in the processes of their emergence in space and time, they are not "nothing but" these processes, but have distinctive ways of acting that emerge through culture. Methodologically, the "naturalism" of the argument is not in the usual claim that some one "natural" science (usually the physics of the day) has a monopoly on methods and a privileged access to truth; it lies, rather, in the insistence that any nonformal mode of inquiry be open to criticism and revision on the basis of some appropriate range of experiences. No revelations, either from Intuition or Pure Reason or God, need apply.

In terms of positions in the great tradition, the argument takes off from Aristotle's categories of potentiality and actuality, and sees the biological and the psychological as the material out of which the distinctly human forms of action emerge. But I see these forms of action as distinctly moral only as they affect the kind of irreducible human dignity that Kant tried to catch in his categorical imperative. Bringing the Aristotelian and Kantian insights together will obviously involve interpretation and emphasis on selective texts and textures in each of their work, rather than a historical attempt to sum up what they each "really said," taken in the light of the concrete historical problems that drove them into thought. My aim is to get at what their ways of thinking suggest for us in *our* historical circumstances and problems. This means, minimally, sending Aristotle to school with Darwin (and how delighted Aristotle would have been with evolution as the intelligible story of emergent forms!).

As for Kant, his analysis must be (with some suggestions from Aristotle) cut loose from the mechanistic rigidity of his Newtonian scientific outlook and made compatible with a position in which time and context are fundamental. Only then will it become clear both how the Kantian analysis of human dignity can be made concrete in terms given by contemporary biological and psychological studies, and how these studies themselves can be structured and enriched by the more formal Kantian categories. To distort a Kantian formulation: Kant without Darwin is empty, Darwin without Kant is blind.

At the core of the analysis is the notion that self and other, organism and environment, are defined and determined in relation to one another. The powers and needs of an organism, for example, are always powers and needs of an organism at a given stage of evolution or development in, and in relation to, a specific environment; and the environment is "an" environment only *for* some organism. In Kant, of course, the relationship between the self and the world as other is stated in nontemporal terms: he is not asking about development and emergence in time and space, but about those structural elements that, in any possible space and time, are presupposed by the actualities of developed self-consciousness. This is a radically different move from that of contemporary developmental studies, where temporal and spatial processes are at the heart of the mutual emergence of self and other in some specific niche. Nevertheless, both Kant and the developmental thinkers share the recognition that subject and object are defined and determined in relation to one another; prior to both is the notion of the process (whether logical or temporal) within which they are distinguished.

Such a wide-ranging exploration will inevitably involve much jumping about from one discipline to another, as well as contentious interpretations of cases taken from history and contemporary life. There is a certain presumption, even *hubris,* involved here, and I can only plead the necessities imposed by scope to ask the reader's indulgence for inevitable lapses from scholarly detail, especially on issues of textual interpretation. Kant, for example, has at least three roles in the discussions. First, a number of aspects of his argument are limited by the scientific thinking of his time, especially the mechanistic causality of Newton and the absence of any doctrine of evolution; plugging in contemporary science changes the arguments. Second, despite the fact that his original argument reflects scientific notions no longer defensible, its formal structure is so powerful and inclusive that it can not only accommodate but also illuminate the new scientific positions. This is particularly true for his argument that self and other are defined in terms of each other, his analysis of the categorical imperative, and his analysis of the exemplary aspects of judgment. Third, if I am to talk about any aspect of Kant without drowning in centuries of technical controversy, I have to take what is at times a rough and ready reading; I'm quite willing for Kant scholars to think of the result as a "Kant-like" (rather than Kantian) position if they wish.

I am aiming at a discussion accessible to the educated reader, not just the specialist, so the intertextual citations and appeals to authority have been kept to an absolute minimum: given the fact that the questions are central to the task of making sense out of human existence, the suggestive parallels and possible comparisons from other fields, let alone the philosophic tradition, are endless and bewildering (my first version of the manuscript resembled those annual Christmas poems Roger Angell writes for the *New Yorker,* where he aims at mentioning the maximum number of names per line). My hope is that readers will find many of the points reassuringly familiar (compared to *X,* paralleled by *Y,* suggestive of the points made by *Z,* perfectly obvious, once put this way—), for the ultimate test of the argument is its power to pull together and give significance to what we know already in all the senses (scientifically, morally, and aesthetically) in which knowing occurs.[1]

Chapter 1

On Not Reducing Agents to Organisms

> There are many strange things,
> and none is stranger than man.
>
> —Sophocles

People and Process

People themselves are the best illustration of process—both of the processes of evolution through which the species emerges and of the developmental processes through which the adult replaces the child. While most educated Westerners agree that both of these processes do occur, the ways of thinking they call for have simply not penetrated our culture or consciousness. Not only do we persist in thinking of things as entities isolated from each other and from time, and aspire to measure them against atemporal standards, but we think of ourselves as another kind of entity among these others.

Think for a moment about the ways in which we talk about ordinary matters. Whether the topic is family or capitalism, the focus is on the thing itself, taken apart from the time and the changing context within which it comes to be. So one hears talk of capitalism as the best system, the only way (no matter what society or context) of maximizing the (fixed) needs of human nature. Or some form of the family is the "only normal" way in which love and caring are expressed, and its static form is commanded by God, or biology, or common sense.

Given the tenor of this talk, it strikes an odd note to ask what the history of these institutions suggests about how their different forms may have emerged out of different processes in response to different problems in other times. So we are unlikely to ask what problems the institution of the family arose to meet: did the demands of a hunting and gathering people differ from what was later needed when sustenance farming (or an industrial wage economy) restructured the situation? Do different cultural forms (including the family) and different natural resources call for different economic strategies to realize the same rough ends (keeping in mind, of course, that ends can be redefined when reached by different means)? Are the forms of institutions—such as gender roles in the family—a response to a situation that no longer exists, such as a different economic situation or the absence of effective birth control? Or has the institution itself changed the significance of the environment within which it is evolving? For instance, has the change in the family from an extended to a nuclear structure changed the meaning of the economic environment as a source of resources and challenges for the family in its new form? The same issues arise when we ask what sorts of capacities individuals *ought* to develop, but neglect to ask whether these powers are to be actualized in a hunting and gathering tribe or in a Wall Street brokerage.

Although we cannot answer these questions with as much precision as we would like, asking them is a minimal sign of taking process seriously. And just asking them eliminates many of the pseudo answers called out by static and ill-formed questions. Unfortunately, when we do take process seriously, we are likely to do so not by allowing the categories of time and change to permeate our thinking and reformulate our questions, but by subsuming any undeniable instances of emergence and change under ahistoric and fixed categories. Thus when we are forced to confront our own evolving animal nature, the result is liable to be either a quick flight into some (nonevolving) realm of transcendence where this animal nature is left behind, or a pliant, postmodern surrender that leaves all judgments equal: "So, of course, things change and it is all relative, so you can't compare, can't judge. Everything is what it has become, and that's that." (Those who have much to do with undergraduates will shudder at the blithe way this conclusion is taken as self-evident.) In both cases the picture of humans as caught up in and defined through time is taken to threaten the possibility of a coherent and value-structured world. Richard Bernstein calls this the Cartesian Anxiety: "*Either* there is some support for our being, a fixed foundation for our knowledge, *or* we cannot escape the forces of darkness that envelop us with madness, with intellectual and moral chaos" (1986, p. 10). One of the results of this anxiety is the resurgence of religious fundamentalisms in all parts of the world, but similar longings for doubt-free foundations can be found in every realm of political speculation, and even in a certain methodological wistfulness that hovers over the various academic disciplines, especially in their denials that their procedures involve evaluative judgments.

One would expect that the life sciences would be the exception to all this skittishness about distinctive forms coming to be under differing circumstances in time. But even they have been reluctant to make the concepts of emergent form and organic functioning central for the analysis of human beings. This reluctance is not based on the fear (so prevalent in conservative religious thought and popular culture) that these categories will miss the distinctiveness of humans, but that recognizing any such distinctiveness would put humans outside the scope of science; it is as if the recognition of a distinctive form involved some ghostly essence too fine for scientific concepts to catch. So "respectable" science has tended to treat those terms that mark the distinctive ways of being human as merely shorthand abbreviations awaiting translation into the more ultimate script of DNA. The concepts of form and functioning seem shot through with perspective, too variable and qualitative to be "real" science; even if nineteenth-century reductionism is officially dead, we lust after its certainties in our epistemic heart of hearts. Unfortunately, this worship of the immaculate conception of facts not only limits the subject matter and methodology of science, but makes the emergence of value in the world into a mystery of virgin birth: people as ethical agents can have no natural ancestors in a world so conceived. As John Dewey says, "The self becomes not merely a pilgrim but an unnaturalized and unnaturalizable alien in the world" (1958, p. 24).

This view of humans as alien in the world of natural processes is also reflected in the stubborn dualism of our ordinary language, where the mental and moral terms still have an honorific echo, but the causal and material terms have all the bite. How much of this is due to lingering religious notions of the immortal destiny of humans alone among all the animals, and how much is uncritical respect for the quantitative categories of "real" sciences, is hard to say. In any case, if we make process the fundamental concept, then both self and other can be seen as emerging and evolving in relationship to one another, not as contrasting examples of different kinds of beings whose relationships are inherently contingent and problematic. This allows us to move beyond the notion that the distinctive qualities of selves, such as the autonomy and responsibility spoken of in question 1, are threatened by their embeddedness in nature.

Twentieth-century developments in science have shown that reductionist and mechanical models are simply not adequate, especially in the biological sciences, our most immediate concern. (I'll expand on this point in chapters 2 and 7, and chapter 6 will extend the argument beyond the biological sciences.) Rather than plunging into an overwhelming literature to make the negative point that mechanical models fail, I will focus on the positive argument that making developmental ways of thinking fundamental does enable us to make sense of the moral (and the knowing and aesthetically aware) person as an emergent within evolution, rather than a ghostly impostor from without. Without this recognition, our sciences will fail to grasp who we are as subjects—and we as subjects will fail to grasp the ways in which science can shape our lives with intelligence. As it is, our power to destroy each other and our environment radically outpaces our understanding of how we can live and flourish together on a fragile planet.

Levels of Process

We can make sense of persons as distinctive, and yet embedded in nature, only if humanity is an emergent form within nature. From an evolutionary point of view, this is both central and obvious. After all, as G. H. Mead points out about Darwin's book, the word "species" is simply the Latin word for "form," and what we face is the task of dealing with the origin of forms (1956, pp. 7–8). In this sense, insofar as a process is characterized by form it is a *system*. In Paul Weiss's terms, this means that there are at least some fundamental processes and behaviors that cannot be understood as "group performances in which the courses of all components are rigidly predetermined in such a manner as to yield automatically a collective product of predesigned orderliness; this would connote 'macrodeterminacy through microdeterminacy,' characterizing mechanisms in contradistinction to systems" (1973, p. 58). Here a system is characterized precisely by its "essential invariance beyond the much more variant flux and fluxations of its elements or constituents" (p. 41). This language about the way in which form characterizes process is not even in principle translatable into a talk about ultimate and isolatable particulars.

But if we must take form nonreductively at each level of phylogenesis and ontogenesis, then it is essential that our language reflect the specific web of elements and relations characterizing each level. It is disastrous to talk about any level of form in terms only appropriate to some other level—usually a level further down the scale of complexity, such as that of the substrate out of which the form emerges. What results is a language systematically barred from getting at whatever is distinctive about the emergent level of form. Thus one of the things distinctive of humans—and one that depends on, or has implications for, all the other ways they are distinct—is that they can act as moral agents. So any language, however explanatory and nonevaluative its intent, which leaves this out will weave an explanatory net through which humans themselves will slip unnoticed. But the animal can be snared in the conceptual net, for the emergence of humans as agents is simply the emergence of one more form in the process of evolution. In terms of significance for us, this emergence is supremely and irreplaceably important; but in terms of the kinds of thinking through which distinctively human moral agency can be understood scientifically, it involves no radical jump from reason to "intuition," from evidence to speculation, or from responsibility to caprice. Methodologically, it is no big deal.

In Aristotelian language, ethical actions give form to a situation for which the nonethical aspects of world and self are matter. For example, the genus of moral virtue is a state of character acquired as habit, a second nature, through doing the relevant actions; this means that the virtues practiced in any community can be understood only through those processes by which specific habits are established under certain conditions, but not others. But the emergence of different patterns of behavior will not be intelligible unless they are seen as the habits *of* an organism with specific powers and needs. And the latter will not be clear apart from (what we now would call) neural structures, and so on. It is not that knowledge of one level allows us to deduce those properties that are, from another perspective, a further and distinctive way of functioning: this would be a case of what Aristotle calls "mechanical" rather than "hypothetical" necessity (*Physics* 2.9). On the contrary, it means that the limits of this further way of functioning are set by that out of which it emerges.

Consider a visual example of this relationship: the most exhaustive account of input-output at the level of nerve stimulation in the eye will not yield a sufficient account of how the eye picks up form and information, or classifies invariance patterns in the ambient array as "objects." But, equally, no account of the latter would be acceptable that did not present the full phenomenology of seeing as a determinate way in which, under limiting conditions, the potentialities of the organism with a certain history and a certain nerve structure were realized (Gibson 1966, chap. 3). Similarly, the most exhaustive account from biology and cognitive psychology of the processes of memory retention, imprinting, and conditioning will not yield even a *description* of habits in terms of their moral status as virtues, let alone furnish criteria for settling moral choices between them. Yet it is equally empty to prescribe courses of action aimed at the inculcation of rules into character, or to punish or forgive those of deficient character as if they did not acquire character under empirical constraints and conditions. So if biology were other than it is, it might make sense to hold six-month-olds responsible for learning habits of continence, but since neurological development has not yet given them control of sphincter muscles, it is simply silly. Yet, in principle, this is no different from the other questions about how capacities limit or facilitate behavior that we find at every developmental and evolutionary stage. If, as Aristotle says, "the affections of soul are enmattered formulable essences" (*De Anima* 403a24), then we cannot avoid the question, Just what *is* the relationship between chemical imbalance in the body and being carried away by an "irresistible impulse"? What is the consequence of oxygen deprivation of the fetus for learning disabilities? And what are the physical or psychological conditions relevant to learning *moral* rules? In general, what are the conditions under which someone can "see the world as" other than his culture and training prescribe (Wittgenstein 1953, p. 388e), shake himself loose from a closed world and ascend from Plato's Cave into autonomy and responsibility?

These are tough questions at best. But they are simply impossible if raised in the dichotomous terms of nature versus nurture, genes versus socialization (see chapter 7 for an extended discussion of this point). At this stage in the argument, though, I want to put things in the most familiarly acceptable terms, and that means language that has at least echoes of the dichotomies that eventually (*pace* Hegel) must be overcome. For if the argument for not reducing agents to organisms can be made even in these implicitly hostile terms, it will be even stronger when eventually recast in terms within which entities at all levels are seen as inseparable from the dynamic structure of the processes out of which they emerge.

With these qualifications in mind, let us turn to the ways in which selves emerge in the process of interaction as distinctive forms. To make sense of such selves, we will need an account of the cooperation of processes through which basic genetic potential is actualized in specific structures and behaviors, of how the genetically coded fetus in the intrauterine environment enters the process through which it becomes an individual. Although it is difficult to say precisely what is "wired in" (a dangerous but preliminary notion) to newborns, it is clear enough that they do not come into the world naked of tendencies and schemas of behavior. Yet all these built-in action patterns, like grasping or sucking, or a tendency to fixate on certain visual patterns, must be coordinated and shaped in experience before the infant can be aware of a world, that is, have a sense of an object or a person—herself or another—as enduring through time. The notion of an object enduring through time presupposes the idea that its earlier states are connected either through persistence or lawful change to its later states; hence the infant can reach the notion of such objects only so far as she builds up the habit of treating them as graspable, suckable, seeable, that is, as being a set of dispositions that can be counted on in anticipatable and memorable ways. But it is equally clear that the infant builds up her own self as lasting through time only insofar as she has a sense of past and future, the very pasts and futures that are articulated through her dealings with enduring objects. She can come to a sense of her own relatively enduring selfhood only within a matrix of relatively enduring others, and in this sense the shape of self and other are reflections of one another (Bowlby 1988; Pipp, Eastwood, and Brown 1993).

Let me put this in a more linguistic way: after Kant (1787) and Strawson (1959) and Kuhn (1970), it is a commonplace that the central terms of any organizing framework are intelligible only in relation to one another. What it means to be "an object" is inseparable from the kind of space and time through which the object is thought to move, and from the kind of causality that governs its changes and interactions as it moves. So it is simply nonsensical to talk of Newtonian objects, lumps of matter characterized precisely by their indifference to change in place and time, as moving through an Einsteinian space/time defined by reference to mass and velocity. The same is true for the notion of self. Different meanings of "self" (or "organism") will be correlated with, and unintelligible apart from, different meanings of "space," "time," "object," "other," "action," and so on. When Wittgenstein, talking about language games and chess games, says that "the meaning of the piece is its role in the game," he is reminding us that the meaning of the piece is bound up with the moves it can make in the spaces of a specifically ordered board: changing the board changes the meaning of the piece that moves over it; eliminating the piece changes the meaning of all the other pieces and of the world through which they move (Wittgenstein 1953, pars. 563, 31). Analogously, what we mean by the "self" of the infant at one week is inseparable from the world through which it moves, and from the rules, here still essentially biological rather than social, which govern that movement. And so on for each level of self-in-a-world, for each complex set of meanings of "space," "time," and "action." There is even a fair agreement in the literature about what is involved at each level—but only until we reach the emergence of the ethical self. For most of us balk at the proposal that the person as self-determining agent is simply a further form that emerges out of, and changes the significance of, what went before.

Consider the early and less controversial stages: the newborn, already the emergent of even earlier interactions between her genetic code and the intrauterine environment, is the material out of which any further developments become actual. We are not talking—a parody of Skinner (1953) aside—about a pure potentiality that can be shaped in any random way, but about a material already having a certain form, specific powers, and specific lacks. This holds whether we are thinking about the lack and limit inherent in finitude itself, that is, in the possession of some specific organ or genetic program, or about pathological detriments such as Down's syndrome. Negatively, these limits determine what cannot come to be, for there are certain actualities, certain forms, that a biological or a social substrate cannot take on. The material *under*determines what comes to be, that is, the kind of newborn the fetus becomes, the kind of person the child becomes. To understand, for instance, how the child comes to be as one who organizes a world of enduring objects under rules, we must go beyond the genetic and biological and look to those cultural and linguistic factors that differentially develop the same biological substrate in humanly significant ways.

What this means for the ethical becomes clear if we consider what is involved in achieving a selfhood defined not merely in terms of a world of persistent objects, but of persistent other selves. I can come to be someone who acts in and moves through a world of other people only to the extent that they exist for me as calculable, manipulatable entities. Only then is there "a world" within which the remembered past is intelligibly connected to the present of my activities and the futures at which they aim. It is a significant point in the life of a child when she cries out in anticipation of her mother leaving rather than in response to some specific stimulation: only then is the other beginning to exist for the child as a continuous source of nurture and affection. It is in terms of this continuity that the child simultaneously comes to be as one who remembers past events and seeks future satisfactions; her persistence through social time makes no sense apart from the patterns of behavior, the rules and roles that make up the institutions of feeding and caring. The principle involved here is that the child can come to be a calculating and prudential being only insofar as she recognizes others as making up a world within which calculation is itself possible.

Not much of this is controversial. But the logic of the developmental process extends further: these self-conscious and enduring prudential individuals can become moral agents who act responsibly and with authority, that is, *persons,* only insofar as they recognize others not merely as means, animate objects who can be used, but as other persons. This is an old theme: Kant put the conditions for entering into this world as an elegant imperative, "Act so as to treat humanity, in your own person as well, always as an end, and never as a means only" (1785, p. 429). To which Martin Buber added that the "I" of personhood is always spoken with the "Thou" of personhood (1970, chap. 1).

The Emergence of Persons within Process

Any scientifically grounded language not only can but must take account of this dimension of experience. As an emergent level of form, it involves the same sorts of structural elements as the earlier stages, that is, the taking on of a new dimension of systematic organization. And structure comes about here, as there, through a redefinition of the relation of the self to the other, to the world that is codeterminate with the self. Thus recognizing others as centers of dignity and self-determination means recognizing that, simply as being who they are, persons set limits on my use of them as objects to be subjected to my will (Kant 1785, p. 428). In this way they constitute, together with the institutions through which they are articulated, a world defining an ethical level of space and time and lawful regularity, a matrix within which my own existence as ethical becomes intelligible.[1]

Consider, for instance, the institution of promise-keeping. At the pre-ethical level I can use the speech act of promising as an instrument to manipulate someone as an object of my will: I promise to do *X* if she will do *Y* first, but I keep the promise only if it suits my advantage. This could include all sorts of "enlightened" calculations of long-range interest (if I don't pay off now, I won't be trusted in the future), but essentially the other is, in Kant's terms, a "means only" to my interest, and not an "end" in herself. But if I were to keep the promise not simply under the hypothesis that doing so would serve my interest, but out of respect for myself and the person to whom I have promised, then I have changed the shape of the world I live in, and of what it means to be myself in it.

Forget ghost tales of noumenal selves and transcendent egos, let alone eternal and noncorporeal souls, and consider what this means in developmental terms. There are trivial promises ("I'll have the exams marked Monday") as well as fundamental ones involving my whole being. To break the latter, to fail a child whose care is my own, is to become a different person. For insofar as my own continuity is determined by a constancy of rule-governed behavior toward an other, the constancy is broken. Compare: only insofar as I can determine and recognize a world of persistent objects can I come to be as a persistent self-who-perceives; only insofar as I can determine and recognize others as centers of dignity and agency can I come to be as an agent who has persistent identity as an agent in a world of agents.

Just as the meanings of fundamental concepts are systematically related to each other within some paradigm, so the meanings of "self," like the meanings of a piece in chess, will be displayed in and constituted by the rule-governed moves on a certain board. It is not pushing the metaphor to say that the spaces and order of the board for the moral self are laid out in the institutions such as promise-keeping, family roles, citizenship, the requirements of informed consent and the doctor-patient relationship (see chapter 3) that define the world within which choices are made at a given point and place in history. To be related to one child as father, and to others as uncle or friend or fellow citizen, or simply as fellow human being, defines different relationships of distance and proximity in moral space. Whatever my relations to an unknown child are as a fellow person, the space of my action is more strictly bound and more finely worked out by my relation to my own child (Jane Austen is as perceptive about the texture of such things as anyone who has ever written). Similarly, in terms of time, my past as a moral being is not fully separable from my role as a father, nor can any of my futures be consistent with my present self unless they are consistent with this commitment. To the extent that this relationship is constitutive, decisions that change it will be, literally, existential. Since they define a new level of agency and development, they must be accounted for both in ethical evaluation and in scientific explanation.

There are, of course, no a priori grounds on which one can give a general and yet substantive account of what it means to have a caring relationship to a child, to treat her as an end and not a means only. This point is quite general, for the structural principle that says one comes to be a specific kind of agent only by treating others as agents in specific ways is a *formal* principle. This is not to say it is "empty" in a pejorative sense, but that it is "procedural" rather than "substantive" (Aiken 1962, pp. 81–82), for it cannot by itself specify in all times and places what counts as treating another as a person—any more, one might add, than the formal principle that all levels of self involve both nature and nurture can by itself predict the course of ontogenesis or phylogenesis. But it does tell us how to *proceed,* what questions to ask: if you want to understand any particular form of self, ask not only about the substrate out of which it emerges, but about the ways in which its structure is codeterminate with the structure of its world. We must ask what components with what history make some salient and important kind of experience possible, for it is only in this transaction within which their powers are actualized that the significance of either self or other is specified.[2]

Methodological Problems and Prospects

Qucstions about the stages of self can arise in quite different problematic situations, since we might (nonexhaustively) be interested either in scientific or ethical understanding, and might contemplate intervention or action in either instance. Furthermore, two activities fitting the same description (like "paying a debt") can have different significances if they come out of different histories: paying a debt to keep my credit rating has a different meaning than doing so out of respect for the person to whom I promised, and out of respect for myself as a promise-keeper. But in all these situations we must be clear about the emergent nature of the self under study or evaluation, for only then will we be able to ask questions—either scientific or ethical—at the appropriate level.

This can be illustrated in a line of reasoning about events at quite a different level in the evolutionary process. In his studies of adaptive responses of the Black-headed gull, Niko Tinbergen was puzzled by the fact that during periods of heavy predator pressure the gulls leave their broods exposed while they remove empty shells from the nest; doing so leads to some immediate losses in the brood, but is functional in terms of species survival. He notes:

> The study of egg shell removal illustrates not only how functional studies guide research in causation, but also that the two approaches are mutually inspiring. The discovery that failure to remove the empty egg shells was penalized by predators made us turn to the stimuli that control the gulls' responses to egg shells. We found . . . that the gulls distinguished an egg shell from an intact egg . . . mainly by its showing a thin edge. This opened our eyes to a new functional problem: what prevents the gulls from removing a newly hatched chick together with its egg shell when it is still half inside. . . . It is . . . this continuous alteration between the search for survival value and the analysis of mechanisms that gives biology, at every level of integration, its particular flavor. The discovery of each particular achievement inspires one to find out "how it is done"; conversely, the student of mechanisms derives satisfaction from understanding how the achievements of these mechanisms contribute to the animal's success. (Tinbergen 1973, pp. 210–11)[3]

The "achievement" that concerns us is the emergence in the developmental process of morality as a distinct stage, one characterized by problems that no amount of science or prudential calculation (however necessary) will be sufficient to resolve. As at any other stage in the process of evolution and development, we must be open to the possibility that understanding the "mechanisms" involved in this new form may entail new modes of inquiry. Just as the analysis of psychological characteristics uses different variables than the analysis of cell mutation, so trying to make sense of an agent acting out of respect, not calculation, calls for categories of understanding neither appropriate nor needed at other levels. "New" methodology need not mean "nonscientific," any more than "unique" means "apart from the developmental process." That there is a level of self for which moral reasons—evaluated in their own terms and with morally relevant criteria—are operative in determining the choices of the self does not mean that these reasons are nothing more than elements within causal sequences understood at a nonmoral level. Quite the contrary: unless one sees morality as an emergent achievement, the scientific analysis (while reassuringly reductionistic) will miss the point.

But why is this so? Would it not be more parsimonious to say that there are some misguided people (about whose peculiarities the Freudians may enlighten us) who take morality seriously and for whom moral reasons "have weight" (Toulmin 1970, p. 17), then interpret this weight as the burden of custom and superstition, not the command of reason? We could then give a straightforward, single-level causal account of how these reasons caused the misguided idealist's actions. For surely empty or even contradictory "reasons" are as causally efficacious as any (see, for example, the history of political decision-making). This sort of argument is behind the widespread assumption that we can do the scientific job by understanding morality in psychological and sociological terms without worrying scientifically (however much one worries personally) whether morality "makes sense" or not.

But consider the parallel move in a nonmoral context. If we take the view of factual statements that social scientists often take of evaluative statements, we would not take them to be either true or false, but simply expressions of the ways in which people of a certain group are conditioned to respond to "evidence." Here there would be no question of sound or unsound reasoning, only the fact of belief, however caused. But as Lawrence Kohlberg points out, "a psychologist or social historian who explained the development of Darwin's belief in evolution in the same terms as an Anglican bishop's belief in divine creation would simply be a poor social scientist" (1981, p. 113). Such an "explanation" would leave out what is distinct in Darwin's achievement, falling back instead on exaggerated emphasis on such mechanisms as indoctrination, socialization, and the use of belief as a rationalization of economic and emotional needs. That these have a role in some science (as in some moral belief) is undeniable; but to say this is all that occurs is to miss a distinction between (at least) most science and (at most) some theology. For such an explanation would fail to detect the mechanisms necessary for the appearance of thinking like Darwin's, but not of rote belief in tradition (this example with apologies to Bishop George Berkeley).

Similarly, one who sees morality as "nothing but" a strategy of gene maximization, the rationalization of a too-timid ego, or a propaganda disguise for self-interest (Thrasymachus's "interest of the stronger," *Rep,* bk. 1) will give an explanation of "how it is done" in terms that exclude or distort the mechanisms making morality possible; they will fail to give an account of the rational and affective capacities necessary to achieve a self whose acts are shaped by the recognition of human dignity. Consider again the nonmoral parallel: if we take all true/false claims to be equally groundless, reflecting no difference in analysis of evidence or methods of reasoning, then our search for the ways in which these beliefs come about will not include a story of the gradual acquisition of canons of reasoning, or of separating out wishes from facts (remember Darwin and the bishop). But the history of science, or the biography of an individual scientist, includes more than socialization and indoctrination into arbitrary assumptions sanctioned only by group approval or tradition. Similarly, an adequate account of morality as the recognition of the selfhood of the other must include the ways moral reasons have weight with moral people.

Becoming moral involves participation in what Stephen Toulmin calls " 'the rational arts'—moral reflection, practical deliberation, intellectual calculation—which are inculcated in us through education and experience" (1970, p. 21). Recognizing these features is obviously important for moral reasons; but it is equally true (Toulmin again) that "our capacity to recognize features of the situation in which we have to act as providing 'reasons' for acting in one way or another, and to act accordingly, is something that cannot be ignored, *even in a complete causal account* of our actual behavior" (p. 17). In other words, that persons hold certain things to be moral or immoral is a fact without which the action would not come about; that certain reasons have weight for them is one of the things that causes them to act, and which must be included in any account of their acts (this is just what the sociobiologists leave out). But while holding something to be a moral reason does not come about without conditions, it is not "nothing but" the sum of these conditions: its significance *as choice* is to bring into being (or reaffirm) a distinctive level of agency, a distinctive way of being in a world now defined in significantly different ways.[4]

What, then, is the status of this claim that ethics is irreducible? Are we left with a situation in which some say there is an "achievement" that must be explained, not explained away, but their opponents find nothing unique and nonprudential that needs special explanation? In dealing with this question, we must remember that "achievement" refers to a distinctive and salient mode of experience that has systemic characteristics not reducible to its components, not to a positive evaluation of this experience. Thus the radical evil of Kurtz in Joseph Conrad's *Heart of Darkness* is, alas, as distinctive and defining of humanity as is the production of symphonies or the deeds of saints. So what we are dealing with is not just the idealist's claim that there are great doings in the world that ought to seize our attention and admiration, but the broader claim that some distinctive modes of functioning, whether ethical or biological or aesthetic, demand to be dealt with as emergent and irreducible levels.

The problem is quite general. Consider the argument between the sociobiologists, who claim all behavior can be understood in terms of gene maximization strategies, and those who argue that culture is an independent variable in at least some (mostly human) activities. Marshall Sahlins sees the possibility of a decisive test of the sociobiological thesis: "If kinship is not ordered by individual reproductive behavior, then the project of an encompassing sociobiology collapses" (1977, p. 18). His test case is an analysis of kinship on the atoll of Rangiroa, and it quite convinces me (which is not surprising, given my assessment of the reductionist enterprise in general). But must it convince the convinced sociobiologist? I fear not, for any proposed counterexample can in principle be dissolved into a mixture of more complex calculations (delayed reciprocal altruism), evolutionary lag, or a story "to be continued" with further research.

Similarly for ethical phenomena: any answer to the question "Why should I be moral?" in moral terms ("You should do it because it is your duty") is in some sense circular; any explanation in nonmoral terms ("You should do 'moral acts' because it is in your long-range self interest") simply misses the point that morality consists precisely in doing an action because one ought to do it, and not for some other reason, however compelling (Bradley 1927). And, indeed, centuries of philosophical controversy illustrate the fact that any claim that a particular act was done out of respect rather than self-interest, however enlightened, can be countered with talk of motives more devious or needs more amiable ("She ran into the burning building to save the child because the pain in her life without the child would outweigh the pain and loss of her possible death," and so on).

So how, in general, *does* one argue for the indissolubility of any "particular achievement," given that the concepts making the achievement distinctive are constitutive in one model and do not occur in the other? I think that all we can do (and it is enough) is move back and forth between the functional analysis of the macroscopic behavior and the analysis of the mechanisms that make it possible, and hope that this (in Tinbergen's terms) "continuous alteration" will indeed be "mutually inspiring." If so, then the activities of moral behavior (keeping a promise even when it is not in one's own interest) will send us to the conditions that make it possible (overcoming the egocentric perspective of early childhood) with new questions and insights; perhaps understanding "mechanisms" (bonding between mother and neonate) will suggest new ways of understanding moral phenomena (like trust) that they help to make possible.[5]

Sometimes putative "achievements" are illusory, not in the sense that they don't exist or don't have significance, but that they don't "work out": taking "possession by demons" to be an irreducibly significant level of functioning suggests no fruitful questions about the mechanisms of possession. Furthermore, an analysis of biophysical and psychosocial phenomena seems quite able to account for such macroscopic phenomena without referring to a special level of spooks. Much more controversially, naturalists like John Dewey (1934b) argue that religious experience does not represent a particular achievement in the sense that its understanding requires a recognition of a special level of divine entity. In Dewey's analysis, religious categories do not refer to an emergent level of behavior that cannot be explained through some combination of talk about moral insight, poetic form, and social control. If he is right, then we can make religion intelligible without postulating gods, but cannot make sense of moral behavior without talking about actions between persons who treat each other as ends. It still might be the case (as I believe it is) that religious experience is distinctive and extraordinarily powerful in its modes of integration and symbolization, even if we need not refer to God(s) in understanding it.

So far I have said almost nothing to detail the interactions that show that the "continuous alteration" between moral behavior and mechanisms is indeed "mutually inspiring" in the requisite sense, but there is already a rich body of work on these issues. Thus, if the question is, "What blocks and what facilitates the treatment of others as centers of dignity?" we can look to Plato's argument in the *Republic* that the person of disharmonious soul will be enslaved by a particular passion and unable to know himself or others; or to Aristotle's analysis of character as a condition of moral action; or to Spinoza on "raising passion to status of an idea"; or to Hume on the disinterested spectator; or to Hegel on the master-slave relationship, Marx on alienation, Piaget on the egocentric perspective, Mead on relations to significant others, and on and on. In nonphilosophical fields, the work of therapists and novelists is particularly rich in insights into the ways in which personal relationships can stifle or stimulate the self's peculiar excellences, just as sociologists tell us much about the vagaries and vicissitudes of living in groups—or being cast from or engulfed by them. Nor are geneticists without prophecy on DNA as destiny, or theologians without evidence on the role of hope and the importance of faith. And on and on. The extraordinary ease of multiplying examples will occur to every reader, usually with particular emphasis on the relevance of her own field, whatever (almost) it is. This very ease drives home my point that we know a great deal about these matters; what we need to do is remove the methodological shibboleths that block intelligent integration. It is in this way that I hope the present argument will contribute to the solution of particular problems, not by presenting some "new" concept of an overwhelming and inclusive good that can serve as a criterion for decision.

I will say more about such matters after additional distinctions are in place, but I have concentrated so far on the preliminary job of arguing that there are no methodological oddities in such an inquiry. On the contrary, it simply extends accepted developmental principles so that they apply to a full range of developmental stages. I will go on to argue that making this extension shows us that (to paraphrase Kant) ethics without science is empty, and science without ethics is blind.

Chapter 2

Biological and Ethical Processes

> Justicė is produced in the soul, like health in the body, by establishing the elements concerned in their natural relations of control and subordination, whereas injustice is like disease and means that this natural order is inverted.
>
> —Plato

Health at Different Levels of Process

If the ethical is an emergent from the biological, then all ethical questions are, in this root sense at least, bioethical. I looked at these issues in a general way in the preceding chapter, but now turn to that shifting set of issues ordinarily labeled "bioethical" in the narrow sense. Examining this rat's nest of problems shows how issues of "health" occur at levels ranging from biological organisms whose health is in well-functioning of organ systems to persons whose health lies in all their distinctive ways of functioning, and involves everything from sound judgment to the preservation of autonomy. Ultimately, I believe, being healthy means functioning fully and authentically in those ways that mark out what it means to be human; although this does not occur in isolation from biological health, it is not reducible to it.

The shift in focus from the merely, mostly physical to the more problematic concerns with persons in their full humanity is reflected in the difference between the reductionist terms of nineteenth-century Western medicine and the more inclusive thinking of the late twentieth century. Predictably, this shift has bothered both those who see it as a falling away from sound scientific medicine, and those who see the "intrusion" of science into "humanistic" concerns as reducing agonizing human problems to an imbalance in the endocrine system. Some of the wilder reactions in the latter vein make up the crazy quilt of "New Age" nostrums, many of which define themselves as much by a rejection of science as by their inclusion of nonbiological notions, such as "spirituality." Their rigid exclusion of the biological makes them less a "holistic" alternative to nineteenth-century biological medicine than an embracing of an equally one-sided focus on another aspect of process.

Any adequately inclusive view must deal with all the emerging levels of process, yet be sensitive to the methodological differences demanded by each, for we neglect them at both our physical and moral peril. Although physical health and moral health refer to distinctive levels of emergent complexity, both are organizational concepts. At either level, "health" (like "disease") refers to a mode of integration in which a number of elements—differing substantially in kind as one moves from the biological to the ethical—are given a unity and significance. Since a nicely intermediate range is picked out by the World Health Organization (WHO) definition of health as "a state of complete physical, mental, and social well-being, not merely the absence of disease and infirmity" (WHO 1946), I will focus on that; we begin by looking at the ways in which the familiar shapes of "bioethical" problems raise issues of conceptual integration and organization.

"Health" and "Disease" as Integrative Concepts

Consider abortion and care of the terminally ill, treatment of defective neonates and the dilemmas of informed consent, *in vitro* fertilization and the use of biosurgery, experimentation with children or with animals, and so—endlessly—on. What these questions share is a common theme of the *development of the organism in time:* in each instance the core and organizing question is whether, at any given time or stage, there is a *person* present as an achievement within process.

As in the previous chapter, I mean by "persons" those who bear rights and responsibilities; they are those, as Kant said, who "are called *persons* because their nature already marks them out as ends in themselves—that is, as something which ought not to be used merely as a means—and consequently imposes to that extent a limit on all arbitrary treatment of them (and is an object of reverence)" (1785, p. 428). From this perspective we cannot avoid the question, What is the relationship, over time, of the physical, mental, and social processes blocking or facilitating the emergence of persons? Here we can run through the liturgy: Regarding abortion, when does the fetus become a person? In caring for the terminally ill, when do we no longer have a person but a potential organ farm? Do certain defects in a newborn mean the emergence of a person is impossible, and the organism can be "killed or let die"? When is sufficient personhood present to give informed consent? But if this is to be more than a ritual invocation of the magic word "person," we must specify just how the organism becomes a person. The relevant theme in this account of how biology becomes biography is that of increasing organization within those processes out of which emerge gene and chemical environment, organ and body, animal and niche, self and cultural world.

Consider first the level of organization involved in the WHO's definition of health as "a state of complete physical, mental, and social well-being, not merely the absence of disease and infirmity" (1946). Whatever ambiguities this definition may have, it is clear that the physical, the mental, and the social do not represent three kinds of health associated together only by belonging to the same individual. They are not like three kinds of financial assets that add up to a sum, as it is good to have cash, stocks, and real property, but all too easy to have one without the others. What has become increasingly clear in biofeedback experiments, the study of psychosomatic diseases, and all the integration that goes on under the rubric of "holistic health," is that the focus is on intersecting levels of existence: the physical is the material factor "out of which" (in the Aristotelian sense) the mental and the social come to be. The child with Down's syndrome is negatively determined in regard to certain social and mental potentialities—with *that* biological substrate, these behaviors cannot emerge. Yet equally, the biological underdetermines the possible macrobehaviors: it is the determinable that is given form only under specific circumstances.

Causal efficacy can also run from the emergent state to its substrate; for example, stress generated by a conflict between the individual's goals and the demands of a social (or ethical) role can be as significant at the level of body chemistry as body chemistry can be in blocking or facilitating moods or capacities at the intellectual and social levels. As Aristotle points out,

> While sometimes on the occasion of violent and striking occurrences there is no excitement felt, on others, faint and feeble stimulations produce these emotions, viz., when the body is already in a state of tension resembling its condition when we are angry. Here is an even better case: in the absence of an external cause of terror, we find ourselves experiencing the feeling of a man in terror. (*De Anima* 403a19–23)

All this is a reminder of the obvious, or of what has recently become more obvious.

We can discuss these matters more precisely by making what is both a linguistic simplification and a point about levels of existence: because human mental activity involves language and interaction with others, we can treat mental and social health as belonging to one level, that of "individuals." This does not imply that "individuals," like the mythological beasts of classical liberal "Individualism," exist as isolatable social atoms apart from the communities articulated through language and institutions. Quite the opposite. But we can make a stipulative distinction between individuals and persons, limiting relations that define individuals to those that are, in Kant's sense, merely prudential: actions of individuals are under the law, but not for the sake of the law. Thus individuals can treat each other with as much or as little amiableness and enlightened self-interest as one might imagine, but not as persons whose nature *demands* respect quite apart from all considerations of pleasure or interest. In terms of the WHO definition of health, the health appropriate to social and mental ways of functioning is the health of individuals, but not yet of persons. At this level, the appropriate norms in terms of which health and disease can be evaluated include all sorts of roles and goals: healthy individuals ought to be able to function in a set of roles, complete a series of tasks, or enjoy certain states of mind. But these functions cannot themselves be further evaluated in terms of whether it is morally good that people have these roles, or choiceworthy that they seek these goals.

These latter judgments can be made only at the further level of integration which includes persons. It is essential to see that the move to this level is simply the further qualification of material by form, not a perilous methodological leap between two separate levels of reality. Just as health in the individual is an organization that makes determinate some of the potentialities out of which it emerges, so moral action further determines the potentialities of the organized individual; here the relationship of the moral to the social and mental (the person to the individual) is parallel to that of the individual to the physical, with the emergent state underdetermined by the material conditions, and the material unintelligible in its significance apart from its further actuality. Consequently, it is as futile to define moral good in a way that ignores an individual's prudential and particular goods as it would be to define disease while ignoring bodily states. Furthermore, there is a moral parallel to psychosomatic disease, for moral action has causal consequences for the individual's ability to function at the nonmoral level: seeking ideals makes a difference in how someone defines the world within which she acts. No science will be adequate which leaves these actions out, or ignores either the physical conditions, or states of character, or cultural institutions that block or facilitate them. What we need is an account of how a physical organism becomes a thinking and interacting individual, and how that individual becomes a moral person—in essence, some story of how biology becomes biography.

This involves seeing just how the biological self as determinable is made a determinate moral self through the progressive organization of a "world" defined in prudential and, ultimately, ethical categories. In traditional terms, this process can be described as making a One out of the Many, where the Many and conflicting ends of our manifold desires are subjected to judgment from an increasingly integrated and unified perspective. It is Kant who has most parsimoniously elaborated this notion of the unified perspective embodied in a single principle of judgment, the categorical imperative; so it is in his terms that I phrase the next stage of the argument. I will argue not only that Kant's analysis of the irreducibility of the ethical is needed by any inclusive developmental account, but that a Kant-like position needs the developmental account if it is to make sense of embodied agents who act in time and space.

Morality as an Ordering Principle

We begin by sketching what is involved in the notion of a category as a mode of synthesis and organization, then go on to apply this to the increasing levels of organization involved in the WHO definition of health.

In Kant's language, a concept like "disease" is a rule for the organization of a manifold, that is, a way of grasping *("begriff")* a number of more elemental items and holding them together. In this way of thinking, a disease as "an object" is "that in the concept of which the manifold of a given intuition is united" (Kant 1787, B13). What gets united in the notion of medical disease is some combination of symptoms—pains and secretions, modes of behaving or failing to behave, twitches and thoughts—which *are* symptoms only in relation to some organizing concept of disease.

If this notion of organizing a manifold is going to be useful on all the different levels of process, we must understand "manifold" in a formal and functional rather than a substantive sense. In this sense, the elements of any manifold will consist in *whatever* is necessary to make sense of the integrated level of process that is the focus of the interest, through which the manifold is identified as such (in the terms developed in chapter 6, the manifold is the "locus of significance" of the inquiry). What we must *not* do is assume that it is always the same level of somehow ultimate particulars that get integrated, no matter what the level or focus of investigation. For example, suppose we took the biological elements relevant to health at the organic level and assumed that *all* notions of health referred to the same elements; every kind of functioning, from psychosomatic diseases to genetic coding, would be understood as a combination of the same level of elements that were (apparently) adequate at the organic level. This sort of thinking is familiar from nineteenth-century physical reductionism, and is far from absent in today's discussions of everything from violence to homosexuality.

If, however, we hold tight to the functional sense of the manifold as *whatever* is marked out at some level of process as integrated by some concept, then we will see that categories like health and disease function at markedly different levels of process by holding together and giving significance to quite different sets of elements. If, for example, we consider disease as a category at the level of health marked out by the WHO definition, the elements of the manifold to be "run through and held together" by the organizing form would include physical states of organ systems, psychological capacities, and social roles; here each "element" in this integration is changed in significance by its inclusion.

For example, pain is an element often included in the concept of disease. But before we can even *identify* a particular pain, we must be in a world so organized that there is a continuing self that can "have" the pain again. Some pain, of course, is positively perceived as part of a desirable process (healing, stretching out tight muscles, the sign of peak exertion) and is not part of disease at all. But even if someone is in pain and wants to be out of it, this does not mean she has a "disease." The pain must be seen as part of a natural (not "supernatural") process, yet not due to an accident, nor yet inevitable (like aging), nor part of a proper human function (childbirth, teething). Pain is not only not sufficient for disease, it is not necessary (the "hidden" disease is most frightening of all). Where pain *is* involved, it is a "symptom" only if it is placed in the time of prognosis and etiology, given significance in terms of lawful relation to "a" disease. This involves re-identifying another "case" in which the disease affects similar bodies in similar ways under similar circumstances—and anything that doesn't have (roughly) the same etiology and prognosis under (roughly) the same conditions isn't (even roughly) a case of the same disease.

Another illustration: to say that people have a disease places them in the "sick role," and this act changes the significance of much more than their aches and twitches. In Talcott Parsons's (1951) classic delineation of the role, it is assumed that patients are not responsible for getting ill and cannot get well by an act of will (unlike someone simply stubborn or perverse); hence they are relieved of some or all of their normal roles and responsibilities. Moreover, since they should want to get well as soon as possible, they have an obligation to seek the help of those who can help them to do so. So read, their manifold activities are indeed "run through and held together" in ways that bestow new significance.

Being in the sick role changes behavior, which points to the obvious fact that health and disease (like "good") are not simply descriptive concepts, but also practical ones, with their home in the language game of action. There they are embodied in practical principles, that is, "propositions which contain a general determination of the will, having under it several practical rules" (Kant 1788, p. 18). Such principles can occur at widely different levels of generality, ranging from the specific ("Take this medicine"), to the general ("Change your lifestyle to decrease stress"), to the universal and purely formal command of the categorical imperative that does not apply directly to particular acts at all.

As principles, they all guide action in accordance with some organized view of the world within which certain actions are choiceworthy. Having a principle involves more than reacting to an impulse (compare: more than feeling a pain); it involves weighing the impulse in the scales and placing it in some order of priorities, however limited and contingent, or universal and unavoidable, this ordering might be. Without this, we are at the beck and call of every random pleasure and pain, like Plato's democratic man who, lacking any principle of choice, declares "one appetite as good as another and [says that] . . . all must have equal rights" (*Rep* 561c). Such a "manifold man stuffed with most excellent differences . . . containing within himself the greatest number of patterns of constitutions and qualities" (561e) might happen to pursue a pleasure compatible with health, but he could not, in Kant's sense, take an "interest" in health (Kant 1785, p. 81). He lacks not only the temporally structured world within which he could choose with a "measure of due amount," but also (and these are inseparable) he lacks the kind of self that could "run through and hold together" such a world: I can be conscious of myself as organized only to the extent that I am placed in an organized world. This is one of the fundamental contentions of the *Critique of Pure Reason:*

> The original and necessary consciousness of the identity of the self is thus at the same time a consciousness of an equally necessary unity of the synthesis of all appearances according to concepts, that is, according to rules, which . . . determine an object for their intuition, that is, the concept of something wherein they are necessarily interconnected. (1781, A 108)

The principle holds equally in the *Critique of Practical Reason,* for "the moral law is in fact, a law of causality through freedom and thus a law of the possibility of supersensuous nature, just as the metaphysical law of events in the world of sense was a law of the causality of sensuous nature" (1788, p. 47). As a consequence, the rules through which the moral law operates have an organizing function parallel to that of the categories in the operation of theoretical reason: but

> these rules . . . contribute nothing to the theoretical use of the understanding in bringing the manifold of (sensuous) intuitions under one consciousness a priori, but only to the *a priori subjection of the manifold of desires to the unity of consciousness of a practical reason* commanding in the moral law, i.e., of a pure will. (p. 65, my emphasis)

At the level of sheer impulse, like that of pure sensations (whether these are possible states of affairs, or the hypothetical result of analysis, is not relevant here), one has neither "a world" for action or for knowing, or a self capable of either. If, to put it in Humean terms, impressions did not linger awhile and become ideas, but passed across the mind like shadows on the water, like the perishing particulars of *Theaetetus,* they could never be grasped as such (Kant 1781, A 103).

Putting this in developmental terms brings us to the question, What are the minimal conditions that *any* organism, in *any* ecological niche, must fulfill if it is to learn *anything* about the structure of any possible surround? Clearly it would have to be open to interaction with something other than itself; it could not be a closed system, one of Leibniz's windowless monads. At least some result of some of these transactions would have to be stored; some stored items would have to be retrievable; some retrievable items would have to be sorted as being "the same again" as some others.

Although it is currently fashionable to dismiss Kant's "subjective" account of synthesis in the first edition of the *Critique of Pure Reason* as a piece of a priori psychologizing, the parody of that argument just sketched does give at least some of the absolutely general conditions for any learning by any finite intelligence, and thus, ultimately for self-consciousness. Note that to say that these operations are transcendentally, timelessly necessary is not to say that the operations occur in some nontemporal realm. It simply means that any particular developmental story must include temporal operations of these types, for it is the necessity that is nontemporal, not the operations.

Philosophers have often ignored these (merely) temporal affairs, preferring to do philosophy of mind as if minds had nothing to do with brains, or brains with evolution. In the developmental sciences, though, there is agreement on at least the broad outlines of how the newborn begins to put together a world with herself in it. Although there is fierce debate about just what contributes to the history within which self and other emerge (see chapter 7), there is no doubt that both must *emerge,* for neither is present at the earlier stages of the process. David Elkind summarizes the developmental research this way:

> The major cognitive task of infancy might be regarded as the *conquest of the object.* In the early months of life, the infant deals with objects as if their existence were dependent on their being present in immediate perception. The ego-centrism of this stage corresponds, therefore, to a lack of differentiation between the object and the sense impressions occasioned by it. Toward the end of the first year, however, the infant begins to seek the object even when it is hidden, and thus shows he can now differentiate between the object and the "experience of the object." (1967, p. 1026)

But as Kant saw so clearly, the identification of persistent objects cannot be separated from the persistence of the identifying self: the *kind* of self the child achieves will be inseparable from the order, or lack of it, in the world she comes to live in. So the self-as-perceiver (and actor, lover, appreciator) is coeval with some organization of the world-as-perceived (and acted upon, loved, appreciated).

When infants can consciously identify an object as a source of pleasure or pain—something to be sought or avoided—they have already begun determining themselves to be agents of a certain kind, conscious of themselves as needing and seeking food or affection, acting in a world with such powers and connections as make it possible to pursue assumed ends by chosen means. Such behavior will reflect the implicit practical principles through which will is determined; such agents are less a set of impulses triggered by random stimuli than actors who impose an order of priorities constituting a world within which epistemic and active selves can be (see "The Finite Act as Constitutional" in chapter 4). In Kant's sense, these individuals are beginning to be able to take an "interest" in a world, to weigh apparent goods in the scale of time where present desire has a history. Their actions have a future within which the meaning of action will be further clarified, revised, completed, frustrated.

Health and Morality as Levels of Integration

These considerations show that the WHO definition conceives of health as involving an active policy of self-realization, an ordering of priorities set in terms of an image of what the self can and ought to be. It is an *active* policy, not because it is always conscious, but because health is more than the absence of disease; it is not a neutral state of affairs, but has its significance only within a range of other activities and limits. Even the notion of "disease," admittedly a less integrative and inclusive concept than "health" (let alone "good"), is itself a rule for the imposition of order on at least certain aspects of self and world.

To move from what is involved in recognizing that one has a disease, through developing an integrated notion of physical health, to seeing the health of the organism as the material basis for the integration of the mental and social behaviors that constitute individuality, is to reach that level of rational determination of the will by a practical principle Kant calls the search for happiness, which is "necessarily the desire of every rational but finite being" (1788, p. 25). This involves the rational ordering of the subjective grounds of determination; it is an achievement, not a lucky gift: "Contentment with our existence is not, as it were an inborn possession or a bliss, which would presuppose a consciousness of our self-sufficiency; it is rather a problem imposed upon us by our own finite nature as a being of needs" (p. 25). In more developmental terms, we can say that happiness emerges as a problem within our own particular evolutionary niche; it presents itself to us as part of the territory into which we are "thrown" by our antecedents. This reflects a contingent state of affairs, for we can imagine being free from distress and disease not by action but by luck, by having come to be in a different niche and evolutionary history. But if this were the case, and health were an unnoticeable gift rather than a vulnerable achievement, its significance would be quite different—as, indeed, would the nature of the animal so blessed.

It is just this notion of the body as effortlessly and normally healthy, and of disease as the external and attacking agent, that is implicitly rejected by the WHO definition's claim that health is an active achievement of ordered priorities shaped into a harmonious whole. It is precisely because it *does* make the concept of health so omnivorous that the definition itself has been attacked for "medicalizing" every aspect of experience, in the end giving health professionals a list of tasks demanding omniscience even beyond that claimed by the American Medical Association. A short reply to this challenge is that any definition of health so inclusive of problems must be equally generous in roping in people who can contribute to the solution. This makes health the concern not just of doctors and nurses but of therapists and sociologists, politicians and prison officials, even philosophers.

A longer reply, which is what I am working out here, is to accept the messy impossibility of separating questions of physical well-being from questions of happiness, and to lay out a coherent analysis of happiness within which rational interests can be intelligently balanced and assessed. The problem is that happiness so often seems so circumstantial and subjective that we (especially those trained in medical science) fear that its inclusion in matters of health will bring on the pollution of the merely arbitrary. So it is tempting to move in the opposite direction and limit our health interventions to the patient as organ system. For *this* aspect of the patient seems to be at least, and possibly at most, roughly the same across cultures, thus reassuringly "objective."

There is a sense in which even Kant would agree with this diagnosis: when practical will goes beyond impulse, as it must, and seeks to harmonize desires, it can rely on intersubjectively valid "technical" principles that relate means to ends. But this intersubjectivity holds only so long as the ends are constant, and ultimately fails, since ends depend upon antecedent and differing "subjective grounds of determination." As a result, the overall principles of integration "can and must be very different in different men" (1788, p. 25). Although we have come a long way from the heteronomous multiplicity of the self of impulse, nevertheless, as long as the organizing principle of the will is "determined by the sensation of agreeableness which the subject expects from the actual existence of the object" (p. 22), the subject will be as manifold as the objects that prove agreeable. True enough, this is far from mere chaos, since (although this is not how Kant would put it) evolutionary pressures will insure a certain compatibility among desires as a simple condition of biological functioning. But biological potential can be actualized in many forms, and the cultures and decisions that make the determinable determinate will introduce their own variations. There is only one alternative:

> The principle of one's own happiness, however much reason and understanding may be used in it, contains no other determinants for the will than those which belong to the lower faculty of desire. Either, then, no higher faculty of desire exists, or else pure reason must of itself be practical, i.e., it must be able to determine the will by the mere form of the practical rule without presupposing any feeling or consequently any idea of the pleasant or of the unpleasant as the matter of the faculty of desire and as the empirical condition of its principles. (1788, p. 25)

In the absence of an order achieved by pure reason as practical, even a *de facto* compatibility between desires would be contingent, precarious, and unprincipled. The fact, if it were a fact, that the various activities necessary to achieve my health and happiness were compatible with each other, or with those of some other individual, would imply nothing about the rational choiceworthiness of such a combination. A particularly unfortunate set of circumstances could make it possible for a set of utterly corrupt and perverted ends to be mutually met (science fiction gives us imaginative examples of this). But Kant's point is that without an autonomous and nonderivative principle of judgment that is not merely an instrument in the service of antecedent desires, we can neither adjudicate conflict or judge its accidental absence: "Subordinate to reason as the higher faculty of desire, is the pathologically determinable faculty of desire, the latter being really and in kind different from the former, so that even the slightest admixture of its impulses impairs the strength and superiority of reason" (1788, p. 25).

This notion that the reason determining action in accordance with the moral law is "different in kind" is at once the source of one of Kant's greatest insights and most intractable problems. The insight is in the clarity with which he puts the ineradicable difference between being determined to do an act because it serves one's interests and doing it because it is right, between being happy and being worthy of being happy. The difficulty lies in the fact that he thought it necessary to banish the moral self from the space and time of the Newtonian world of mechanistic causality within which free action was impossible. This saved the purity of the will, but at the cost of making it a timeless ghost in the world of action; it left us with the need for some noncausal pineal gland to mediate between the nontemporal will of agents and their own actions. Worse yet, it gave us nontemporal moral concepts that can make no sense of moral history, of growth and decay, of guilt for the past or responsibility for the future. In these concepts we lose the very possibility of act, for acts must aim either at that which is not yet, or at preserving that which would otherwise perish in the flux of time. The notion of a "nontemporal act," whether of God or the noumenal self, is an empty shell. In the end, one can make sense of acts, and of the autonomy and responsibility they generate, only to the extent that they are embedded in histories and biographies within which they are constitutive.

In short, if the rational will becomes a One apart from the Manifold of desires it is to determine, it will be empty, just as the Manifold of desires apart from the rational will is blind. The "formula" solution is to see the rational will as a One *among* the Many, the act that makes the merely potential into an actual unity. This is precisely what is suggested by our biomedical paradigm, namely, that the moral health (using health as a "typic" to give sensuous embodiment to the concept of the moral good) of the total person will be a form that arises out of and gives a new determinate significance to material constituted by the empirical conditions for the satisfaction of desires and the health of the individual.

Emergent Levels of Space and Time

We have seen that achieving a unified self involves organizing a world within which self can coherently integrate the meaning of its past with the demands of the present in terms of the possibilities of the future. The move from impulsive reaction to "taking an interest" involves organizing the manifold so that there is an object of desire in terms of which calculation can take place. This means, minimally, placing that object in some matrix that is an analogue to space and time, and within which it can have an intelligible career.

This parallels the move in the *Critique of Pure Reason* from "seems to be" to "is," the latter involving the concept of an object that "prevents our modes of knowledge from being haphazard and arbitrary" (1781, A104). It is only when the manifold has been so determined that there is an entity (substance) whose state and place changes intelligibly (causality) through space and time that we can discriminate subjective appearance from phenomenal reality. Similarly, to act toward someone as an individual, for example, as a physician, is to see her as having predictable powers and limits in terms of which I can function as a patient, see my own past and future in terms of the changing demands of the sick role, and so on.

It is only marginally metaphorical to say that we pursue these health policies in an "institutional" space and time determined by the expectations and interactions of individuals. Expectation and roles determine the significance and shape of the time in which we act, just as "friend," "spouse," and "acquaintance" place us at different distances from one another. Obligations and commitments, love and indifference, set the distances within which action happens. All these sociological clichés can be summed up by saying that institutions, defined as shared habits of action, determine a matrix within which individuals come to be through acting (this is what J. W. Miller calls the "midworld"; see chapter 4).

Consider, for example, how the institution of promise-keeping defines a space and time for moral action: as a physician I agree with my patient to respect her wishes to be allowed to die without medical intervention. This can only be as a real promise, not just muttering words, if my agency at the time of promising is related lawfully—*through the promise itself*—to my agency in the future time when the promise is to be kept. Who I now am takes on new significance in relation to who I am not yet. This is a temporal relation between present and future that is characteristic of moral time, and unintelligible apart from it.

I am at least partially determined and defined as a moral being through my relations with the person to whom I promise. Just as her agency—who she has to be in order to be the kind of being to whom I can promise—does not rest in the merely accidental fact that she has a certain set of desires, so my agency is not defined by the mere fact that the promise seemed prudential when I made it. To make a "promise" under the condition that I will keep it only if it serves my own interest is to calculate how that individual can serve my own changeable desires, but it is not to treat her as a person to whom obligation remains constant apart from desire.

There are parallels here between knowing and acting: thinking of a physical sensation as a "symptom" changes its significance by placing it in a time-order of other events; thinking of an individual as a patient or physician redefines the significance of her individual properties by placing her within (differing) expectations and priorities; thinking of a person as falling under the moral law (as, for example, existing in the space and time defined by promise-keeping) redefines the significance of her role by seeing it as a role taken by an agent. This means that the sense of her continuing identity that we are concerned with cannot rest simply in the fact that she is a desiring organism, or even in the fact that her desires are ordered by her actions under the maxim of individual health and happiness. On the contrary, her identity in the relevant sense can only rest in the fact that she is an agent capable of judging that even this harmony of ends (a complete state of physical, mental, and social well-being) falls short of being a unified self in one coherent world and time; she must be an agent who can judge the prudential in terms of the moral.

Here, as in the *Critique of Pure Reason,* the relation of unity in the subject is codeterminate with unity in the object, although here we are talking about moral subjects and moral objects. Thus when Kant says that the job of the rules of morality is to bring about "the a priori subjection of the manifold of desires to the unity of consciousness of a practical reason commanding the moral law" (1788, p. 65), this applies as much to the determination of the identity of the other as a person to whom respect is due as it does to the agent herself. The "unity of consciousness" can only be achieved through acting on a practical principle that organizes a world within which just this sort of unified agent can act: "This consciousness of my existence in time is bound up in the way of identity with the consciousness of a relation to something outside me" (1787, Bxl). To think of persons as ethical substances is inseparable from thinking of them as being members of a community within which they move lawfully through a series of reciprocal relationships with other persons.

In terms of Wittgenstein's notion that the "meaning of a piece is its role in the game" (1953, par. 563), the person's meaning as ethical substance is in the moves she can make on the board defined by moral institutions. Just as in chess, changing the shape of the board changes what it means to be a Queen. And conversely, since the significance of the board and of the pieces is codeterminate, changing the moves of the Queen would change the significance of the structure of the board. So if I act on a board that defines the patient-physician relationship as a contract trading self-interest for self-interest, then both physician and patient roles will have only prudential significance. But if the institutional board defines the relationship in terms of unavoidable respect rather than shared interests, then moving on *this* board gives me—whether I am a physician or a patient—a different meaning. For this relationship defines a different future and gives another significance to the past. Breaking the terms of this sort of relationship says something different about who I am. But however much the human significance of persons moving on the moral board may differ from the prudential maneuvers on the board of enlightened egoism, the process at the moral level is no more problematic, no less subject to empirical inquiry and responsible judgment, than that of individuals described in psychological and sociological terms. Compare: I can be an individual who acts on heteronomous motives with calculation and prudence only if there are calculate-able others existing in a world where past instances supply grounds for actions aimed at a specific future interest; I can be a person who acts autonomously out of respect for the moral law only if there are respect-able others, persons, existing in a world where action can interpret the responsibilities of the past in terms of the obligations and goals chosen without regard to interest.

Unfortunately, however, some of what Kant says about this distinctly moral level of action makes it formidably mysterious. He describes the process we have been looking at this way: "This law gives to the sensible world, as sensuous nature (as this concerns rational being), the form of supersensuous nature, without interfering with the mechanism of the former" (1788, p. 44). His phrase "form of supersensuous nature" refers to a world in which respect for another person can be efficacious in shaping action, and I have argued that this can be understood as a level that emerges when persons are related to one another in lawful ways in an ordered world of moral institutions. This does not exclude persons from the space and time of action, but instead works out levels of space and time appropriate to the different levels of form. Otherwise, each level of emergent form will seem equally mysterious, not just the emergence of the moral from the prudential, but the prudential from the biological.

For example, I have been talking in semi-Kantian terms about the prudential self as a further organization of the potentialities of impulse and sensation; but it could be argued that since this prudential self is not reducible to the self of sensations and impulses, it must be (like the moral self in relation to the prudential) "timelessly noumenal" and mysterious. And, indeed, from the point of view of a radical reductionist like B. F. Skinner, "How can the prudential will be practical?" ought to be a real question. But it isn't, because the emergence of new levels can be quite unproblematic.

We must also avoid the mysterious when making sense of Kant's claim that the imposition of the form of the rational will must come about "without interfering with the mechanism" of the world of sensuous nature. Minimally, this means that the price for making sense of the moral order must not be making nonsense out of genetics, evolution, and psychology. But equally, we must not make nonsense of the fact that the moral will, as the "faculty of determining . . . [a person's] causality through the conception of a rule" is effective in the world of sensuous nature. We must take account of the fact that reason "incorruptible and self-constrained, in every action holds up the maxim of the will to the pure will, i.e., to itself regarded as a priori practical; and this it does regardless of what inclination may say to the contrary" (1788, p. 32). Acting morally has to make a *causal* difference.

Consider this: insofar as my identity as a person is articulated through a relationship to another, and that relationship is embodied in an institution (doctor-patient, father-son), then acting to deny the rules of that institution (contradiction in the will) is acting to deny the continuation of myself as that sort of person (there are times, of course, when new forms of personhood *ought* to emerge; see "Moral Judgment as Both Reflective and Determinant" in chapter 3). Here the fact that the contemplated action would contradict the rules of such an institution would "have weight" in determining my will. "Yet," as Stephen Toulmin points out,

> such aspects or "considerations" are capable of carrying weight only because the agent in question has learned to recognize their relevance to the deliberative procedures in which he is engaged. (Only a man who knows how to integrate can recognize the significance of boundary conditions—only a man who understands tax-law will know how to respond to changes in the interest-rates; and likewise for other modes of reasoning.) (1970, p. 16)

Likewise, only a man trained in medicine can see certain physical complexes as symptoms calling for intervention; only an individual who has organized his roles and capacities in an order of priority can take "an interest" in health, thus making the fact that some activity has certain physical, mental, or social consequences a reason for seeking or avoiding it. And only a person who has organized his roles in ways that define him as related through moral institutions to other persons can take these persons as objects of respect. The fact that his action blocks or facilitates the personhood of others will be a reason that has weight only for a person defined in this way. And a person can be defined in this way only in terms of a moral community within which such a definition can be articulated and revised.

Replies to Some Criticisms

The response to this sketch of causal efficacy is likely to be either that it is factually false or trivially true. It looks false if saying there are reasons implies that there cannot be causes, or at least no causes in regard to these particular acts. Or, in the terms that worried Kant: since there are causes for everything in the spatio-temporal world, an essentially spatio-temporal agent cannot act on reasons. Kant's solution (put crudely and unsympathetically, as it so often is) was to speak of reasons as determining the will of the noumenal self, leaving the phenomenal self to be pushed about by causes. I have been trying to avoid this problematic dualism by talking about moral selves as emergent within their own sort of spatial and temporal institutions. In relation to these, the physical substrate—the undisputed home of firing neurons and DNA—is not *incompatible* with agency, but simply insufficient to account for its emergence and functioning. The biological (and social) substrate underdetermine latter stages of actuality, and must be incorporated into further levels of functioning through which simultaneous levels of self-in-a-world are articulated.

Toulmin's analysis of the way in which reasons have weight for an agent makes the same point, namely, "that the giving of reasons in justification of our actions is distinct from, yet compatible with, the discovery of causes in explanation of these actions" (1970, p. 21). He can put the moral agent in the world of cause because what may be " 'assimilated to causes' is not a man's *reasons* for an action, but rather his *having reasons* for that action—i.e., his *recognition* of those reasons as having weight for him in his particular situation" (p. 19). The fact that—right or wrong—the agent recognizes these reasons as having weight is a causal consideration.

In a quite fundamental sense, this story of the emergence of levels of self as codeterminate with different kinds of worlds is simply the story of the kind of agency necessary for certain kinds of considerations to be causally efficacious. Among these considerations is the role that communities have in making possible the different human achievements that make up individual biographies. Only within communities where relations of mutual respect are constitutive can we make sense of Kant's claim that the pure will is practical; only here can persons be moved to action not simply by the belief that an action will make them happy, but by the belief that it is right. Hence even a scientific inquiry aiming at description rather than evaluation must deal with the fact that people confer meaning on themselves and the worlds within which they act.

There are parallels in the notion of psychosomatic disease (for this need to recognize emergent levels). As Rene Dubos points out, contemporary moves beyond the limited doctrine of specific etiology have "opened the scientific analysis of the responses made by man and animals to various situations, not so much by introducing new techniques as by accepting the fact that many aspects of life cannot be studied except when the organism functions as a living entity" (1979, p. 36). This point can be illustrated with the treatment of cardiac problems and their relationship to stress: "For example, to understand certain features of heart disease, we may wish to focus on cell membranes and electrical potentials; but to investigate whether certain personality patterns are correlated with increased incidence of heart disease, we would do better to shift our attention to higher-level systems" (Brody and Sobel 1979, p. 90). In other words, to understand the health situation of an individual in *causal* terms involves not only such things as the timing of the daily pattern of hydrocortisone secretion, but the way in which the patient perceives the significance of her symptoms as they fit into an ordered world of interests and priorities. Thinking about her tension as a sign of approaching heart failure has different consequences than thinking of it as a helpful reminder that she should—and can—take some remedial action. Changing the significance of the tension by incorporating it into a different meaning structure can be as causally efficacious as using drugs. I need hardly add that these meaning structures are not "private languages," but social structures.

The fact that the patient's beliefs about her stress or pain must be taken into account in any analysis of stress situation and body chemistry does not mean either that her beliefs are nothing but a chemical process, or that they could occur *without* chemical processes. What here characterizes the psychosomatic applies as well to the moral-psychological level, for recognition of others is causally efficacious in action not because it is a further desire or heteronomous motivation, but because it changes the significance of the "manifold of desires" by subjecting them to the "unity of consciousness of a practical reason" that defines a new level of emergent self. Consider the parallels: thinking of the pain as part of the regenerative process of the muscles integrates it into an order, gives it a past and a future, which makes it other than it would have been. Similarly, if someone thinks of her approaching death simply as escape from pain, then it has one significance and weight; but it has quite another if, like Socrates, she thinks of it as an affirmation called for by the moral principles that have defined her life. Kant puts the general point this way:

> Since the idea of the moral law deprives self-love of its influence and self-conceit of its delusion, it lessens the obstacle to pure practical reason and produces the idea of the superiority of its objective law to the impulses of sensibility; it increases the *weight of the moral law* by removing, in the judgement of reason, the *counterweight to the moral law* which bears on a will affected by sensibility. (1788, p. 78, my emphasis)

But even if this position is true, is it more than trivial? Does it say more than that people who take morality seriously act differently than those who don't? Even if it is intelligible that they do, is it intelligent? Why become the kind of person who voluntarily restricts her actions by recognizing the rights of others? Why not just talk a good moral game but do what profits? (Compare Plato's discussion in the *Republic* of whether a rational man with the Ring of Gyges would still be just.)

One answer that obviously will not do is the argument that becoming being moral pays off by making one healthier or happier, where both of these notions are defined in nonmoral terms. These alternatives are excluded by the fact that the "respect" one has for personhood as embodying the moral law "determines the will by itself and not in the service of inclinations" (Kant 1788, p. 24). Otherwise, morality would not be a principle of organization through which persons acquire agency and unity, but simply another interest to be adjudicated—in a situation lacking any principle of adjudication.

It is essential to be clear on the context within which this question about the "reality" of morality is being asked. If we take evolution and naturalism seriously, then history is a temporary phenomenon summed up in the Stoic epitaph, "I was not, I was, I am not." Although the moral law defines the ultimate purposes that structure the relations between persons, it itself has no further purpose; it fits into no further history (or so question 3 assumes). And if morality, although necessary for the emergence of persons, is simply an accident in the cosmos, then we must accept the emergence of the ethical accidentally out of the nonethical—where "accidental" means without purpose, not without cause.

Although Kant did not face this problem, its structure is typically Kantian: Given the existence of some form of experience that is taken to be of demanding significance (coherent sense experience, the duration of the self through time, a sense of unconditioned moral obligation, the disinterested power of beauty), what are the conditions without which it could not be? That we are asking, as Kant did not, for the conditions of the emergence of these modes of experience in *time* does not make the question radically different. Thus, as we have already seen, what Kant says about synthesis as a condition for determining an object for cognition also applies to the conditions for the evolution of any organism that can learn from its surround: whatever the chemistry of life, and whatever the shape of the ecological niche, any organism that can become a learner must be open to stimuli, must in some way store, retrieve, and re-identify at least some aspect of some stimuli. In a parallel sense, a condition of the biological individual becoming a moral person is that she organize a world through a principle that recognizes others as persons who possess a dignity, the recognition of which is sufficient to determine action.[1]

Implications of the Bioethical Parallels

At this point we have come full circle to the opening senses of bioethics as a redundancy: The principle of ethical judgment involved in the evaluation of biomedical problems is the same principle involved in reorganizing the prudential world to make possible the emergence of persons. But it is now clear that we cannot ask whether this meaning given to the world constituted by action is found in that world independently of persons acting in these ways. To ask "Why be moral?" from some position outside of morality is an illegitimate and dialectical inquiry in Kant's original sense. That dialectical position consisted in asking if the categories whose use is justified because they are conditions of determining a knowing self in a knowable world apply to the world apart from the way they have made it knowable. The same move in regard to morality would be to ask if the principle—respect for persons—which is the condition for determining a world within which moral action is possible applies to the world apart from the ways the principle has made it a world of action. Put shortly: Apart from all possible action, what is the status of the principle that makes action possible? Compare: What can be known about the world apart from all the categories that make knowledge possible?

The other side of this process of organization is a process of coming to self-consciousness. Only insofar as the self can distinguish itself from the other can it come to consciousness of itself as other than and standing in some relation to its own sensations or impulses. Only then can it distinguish what appears to be (either good or true or beautiful) from what is. Seeing our own acts and thoughts as contingent is a condition of seeing them as subject to rational evaluation and revision; only if they could be other than they are is it intelligible to raise epistemic or evaluative questions (we deliberate, as Aristotle reminds us, only about that which is within our power). Here being a self in a world is fundamental, for only if we have "a place" in the world of others can we "take the attitude of the other" and see our own position (at least somewhat) apart from the distorting lens of ego (Mead 1956). The roles laid out by institutions as shared habits provide just such points of viewing. Sometimes we can literally have the same role as the other, being both a lender and a debtor, a father and a son, an artist and a critic, a boss and a laborer, a subject and a legislator (Piaget has some lovely things to say about how role-switching in children's games can facilitate this adult transcendence of locked-in perspectives [1965]). Even where there is no literal switch of perspectives, concrete institutional roles can be fleshed out with imagination, including that of art as the creation of alternatives to the actual; here the concrete individuality of a way of being in a specific world is brought to immediate intuition. For many moderns, this is one of the fundamental powers of film.

This consciousness of being an agent is not exhausted in the ability to do this or that specific task (although it does not exist apart from such tasks), or even in the systematic integration of all tasks under the maxim of happiness. On the contrary, agency focuses on the power of persons, oneself included, to integrate all tasks and contingencies so that they, and simultaneously oneself, have an order and identity. This involves the projection—with wildly varying degrees of self-consciousness or originality—of some narrative form that "runs through and holds together" all the manifold diversity of our acts and omissions, ages and stages, hopes and defeats. But this is itself incomplete without the ways in which individual biography is shaped by, and shaping of, the biographies of those others with whom we come to be. Persons must be in histories, in the complex communities within which action can be more than impulse, belief more than a guess, and appreciation than a tingle of pleasure.

So we turn now to history as the context within which human significance arises, and fails.

Chapter 3

The Exemplary Status of Moral Acts

> Night. No one prayed, so that the night would
> pass quickly. The stars were only sparks
> of the fire which devoured us. Should that fire
> die out one day, there would be nothing left
> in the sky but dead stars, dead eyes.
>
> —Elie Wiesel

Meaning in History

Of the three questions posed in the introduction, I have been chipping away mainly at the first two, arguing that by taking emergent form as central we can make sense of human action as autonomous and responsible even if the norms on which we act are our own. This involves a shift from individuals acting in a world of resources and threats to persons acting in a world of others whose dignity is codeterminate with their own. Within this emergent level of activity, the focus is on the continual and constitutive relations to others that embed persons in history. Since on my naturalistic assumptions this history is not itself integrated into a further context of meaning, the distinctive meaning of these lives is found in a history that itself has no external meaning. This brings us to the third question of the introduction, "What sort of meaning can individual lives and choices have if the history within which they come to be is simply an accidental episode in an indifferent universe?"

Trying to make sense of our lives and deaths inevitably involves us in talk about the future, but each of us has a few score years at best. Even as a species, our claim on the future is, in any cosmic sense, trivial; if we do not incinerate ourselves with nuclear weapons or choke on our own filth, then star-death and the processes of decay will bring a more "natural" end soon enough. Not only will we ourselves not "go on," but there will be no one to remember us and hold our deeds in mind. These issues of remembrance and meaning are raised by the Holocaust, for this is as close as we have come to the radical evil of nuclear or environmental destruction that leaves no world behind. The survivors of Hitler's death camps rightly demand that the world hear and remember what happened, not only so that it will not happen again (as it already has in the former Yugoslavia), but because unnoticed and unrecorded death seems to diminish those who died.

But would it? Hannah Arendt says that even if the Nazis *had* succeeded in wiping out the memory of their genocide, it would not have made the victims' sacrifices "morally meaningless . . . it would only have been practically useless" (1963, p. 232). She adds that the "holes of oblivion" into which the victims would have disappeared "in silent anonymity" simply do not exist: "There are simply too many people in the world to make oblivion possible. One man will always be left alive to tell the story. Hence, nothing can ever be 'practically useless' in the long run" (pp. 232–33). Even in the short run the present level of nuclear overkill makes nonsense of her claim that "one man will always be left alive to tell the story." And in the long run we're all dead. But this just adds intensity to the question she raises: what *is* the relationship between the "moral meaningfulness" of action and a threatened or inevitable "practical uselessness" in which action falls into a "hole of oblivion"?

History and Evolution as Limits to Moral Meaning

Since the task of working out meaning in the context of ultimate oblivion has always been fundamental from the naturalistic perspective, the increased probability of nuclear or ecological destruction does not in itself raise a new theoretical issue, however much it makes radical response more urgent. Whether at any given point in history the situation of humanity is relatively secure or absolutely perilous does not change the fact that what it means to be human must be worked out within history (possibly including the history of other evolved intelligences) without reference to an ahistoric grounding or an eternal remembering. If history is not *for* anything or any One, then the naturalist has the quite Kantian problem of making good on a notion of morality that must command categorically if it is to command at all: the ultimacy of oblivion rules out even a covert motivation to do one's duty either because God will arrange an afterlife where virtue will not be "practically useless," or because our deeds will echo endlessly in the minds of our posterity.

Both Kant and the naturalists have their difficulties. His is to give an account of moral motivation that does not reduce it to a prudential wager on heavenly reward. At times he despairs of doing this and reluctantly allows that without the postulates of God and immortality our ideas of morality "are indeed objects of approval and admiration, but not springs of purpose and action" (1787, A 813–B 841). At other points he is firm in rejecting—as his account of the categorical imperative requires him to do—the notion of God as paymaster.

That the naturalist does not face Kant's problem of incorporating an ultimate promised reward into moral action restates but hardly reduces the problem. Both need an account of moral imperatives as grounded and fulfilled in the historical and evolutionary time of human activities, even though action in that evolution is not for the sake of a reward outside of it. I believe that a process interpretation of Kant's reasoning gives us just such an account, one that relates moral obligation essentially to the world of action in time. Here the acts making possible the emergence of self and other are organized on a principle of mutual respect. In Kantian terms this would be the decision of the "elective will" (*Willkur*) to organize itself in accordance with respect for the pure commandment of the rational will (*Wille*) through the categorical imperative. In the developmental terms of the preceding chapter, we can say that the self can reach a stage where moral reasons "have weight" with the person.

There are, however, conditions that affect both what achieving such a stage can mean, and our capacities to reach it. Thus moral agency is limited by the fact that no act is the imposition of pure form on pure potentiality. Consider, for example, the choice to become the kind of self for whom moral reasons are compelling and motivating: it cannot be moral reasons qua moral that are compelling in the "decision" to make moral reasons compelling. This is reflected in Aristotle's point that in becoming virtuous by doing virtuous acts, we first do them mechanically, by rote or under compulsion, and only later with a full consciousness of the principle involved (*Nich. Ethics,* bk. 2, chap. 3). And when Plato talks about the event that breaks the fascination with the shadows on the wall of the Cave and first frees the soul to turn upward toward the Real, he also emphasizes the absence of control by reason and will: "Consider, then, what would be the manner of the release and healing from these bonds and this folly if in the course of nature something of this sort should happen to them. When one was freed from his fetters and compelled to stand up suddenly and walk and to lift up his eyes to the light" (*Rep* 516c). So it may be a piece of "moral luck" that one lives in a polis where the "rational arts . . . are inculcated through education and experience," and where the institutions of public and private life lead to such habitual courses of action as could upon reflection be chosen for their own sakes (Toulmin 1970, p. 21). It is, of course, the job of the intelligent statesman (or parent, or teacher) to see that just such conditions of choice, unchosen by those they benefit, are there by choice rather than by chance.

This same puzzling admixture of contingency in act is present in the evolutionary emergence of moral behavior. Whatever causal processes made it possible, and however intelligible or mysterious it may be from the laws of gene selection and organic survival, from the moral point of view it just happened to happen: evolution spawning morality served no purpose when it happened, any more than it will frustrate one when it ends. I mention so obvious an implication of naturalism only because it emphasizes the role that accident and limit can have in the coming into being of morality at any level, whether of species or persons, of families or nations or institutions, and whether in regard to morality's first appearance or later changes. This is true whether the limits within which choices are made (as in choosing lots in the Myth of Er) are set by the circumstances of evolution or by the choices of other agents, by natural necessity or sheer caprice. This makes it less surprising to find that although moral motive is "for itself" if it is to have any authority at all, it no more dominates the results of acting than it controls the conditions for the origins of action itself. Indeed, there are some situations where moral action, no matter how heroic the agent's intent, simply cannot "come off," for circumstances simply will not let the relevant potentialities be realized. This is one of the most horrifying and ultimate threats posed by the "holes of oblivion" opened up by holocaust.

It is a condition of finitude, and hence of action itself, that acts be *under*determined by the material out of which they come, just as they underdetermine that which they initiate. The act intended at the moment of decision is shaped and modified by the world "into" which the agent acts; circumstances can intervene, and the act that would have made its agent a murderer misfires, fails to come off, and the agent—intention and will unchanged—bears moral status and responsibility for attempted murder only. The driver heading home from a party after too many martinis, or with a car late for a brake repair, intends no harm and may be lucky enough to come to none, but a child running into the street may make her guilty of vehicular homicide (Silber 1960, pp. xcv–xcvi, ciii–cvi). A president like Nixon can intend crimes against his opponents but be frustrated by chance and the structures of government, leaving him responsible for less viciousness than he intended. And it may be that a heroic sacrifice is totally unnoticcd, quitc "practically usclcss" as it falls into the "holes of oblivion" without entering into the ongoing world as example and exemplar; it is thereby diminished not just in its usefulness for moral education, but in its moral status as an act, for it is incomplete as act insofar as it cannot enter into the world sustained and proposed by action.

Since acts neither create and dominate the conditions for their own occurrence nor impose a pure form on a passive and purely potential world, they are subject to a kind of natural limitation, a falling away from moral meaning that reflects the limits of will rather than its corruption. History is the same: all our acts together cannot impose the meaning of humanity except partially and temporarily on a universe that is irreducibly not our own. Nature lets us happen, and achievement of distinctive human experience is indeed a real aspect of nature. But this is not the only or the final aspect of it, for nature will ultimately absorb history. This engulfment of the moral by the nonmoral will not be a moral evil, for that can occur only *within* history. But a limit it undoubtedly is, and we ignore it at our moral peril: not only can individual deaths and that of history itself pass out of all remembrance, but our ordinary lives mostly slip away without much fuss, our peculiar stories remembered by friends for a few years and by families for a few more. Our having lived changes the statistics by one unit, but not much else. It is partly to stave off this terror of leaving no trace that we build our lives into myths of blood and destiny, for otherwise we are too brief to be heard.

Nothing is so clear to the old (or so offensive to the young) than the role that chance plays in whatever life one manages to put together. At worst, conditions conspire to make it heroically unlikely that we can make the smallest act our own. This is what is so offensive about poverty when it is more than just the absence of consumer goods: people are poor when circumstances do not permit their acts to build up a world within which a meaningful life can be lived. As "stepmotherly Nature" sets the natural limits of action, human fear and greed are all too likely to narrow them further. To deliberately act so as to deprive others of the possibilities of a world of action and hence of their distinctive humanity, is immoral—and if done for its own sake, radically evil.

Acts in Moral Space and Time

Many of the problems of specifying the moral limits of act revolve around the relationship of acts to time: if acts were instantaneous effects of will, timeless bursts of noumenal energy and intelligence, then it would be plausible to say that intention is all that counts. In that case, not only would it be "impossible to conceive anything at all in the world, or even out of it, which can be taken as good without qualification, except a good will" (Kant 1785, p. 387), but this will would be the only thing to be taken account of in judging the agent. But acts not only occur at a time, but shape and are shaped by time. Thus we must consider the material out of which the act comes, the efficient factors (in addition to the will of the agent) by which it takes the form and significance it does and has the final outcome it reaches (whether intended or not). For an act is the making actual of what was an underdetermined potentiality (not indeterminate, *pace* Sartre). It takes on its meaning not in some instantaneous moment of intent, but as it grows out of the past, expresses a steady state of character and resolution, and completes that which has gone before (Aristotle, *Nich. Ethics* 1105a31). Any such completion will either modify or confirm the significance of the past by making it now lead to this future.

This point recalls our earlier discussion of the ways in which actors are defined through their moves on the board of institutionalized actions—a board that is itself redefined by these same actions as they impose new significance on the past and invoke a different future. In this process of self and world articulation, we can distinguish three modalities in the relation between the self and the principles that guide its acts. Thesis: the singular act embodies a maxim that, when generalized, proclaims a rule of action that would change the shape of the institutionalized world within which the self acts. Antithesis: a rule grounded in the past and reflected in the institutions that define the existent moral world guides the act so that the self is shaped to conform to the rule. Roughly put, the movement emphasized in the thesis is that of the thrust of the self into the world, that of the antithesis the thrust of the world onto the self. Synthesis: here there are two forms, the first of which is more stasis than synthesis, for it is a mechanical combination of the thesis and antithesis where both rule and self are fixed, yet so related to one another that neither demands a change in the other. The reasons may be choice or chance, dullness or denial, even a pathological condition in which both self and rule are corrupt. It is enough that, for the moment at least, their relationship does not demand revision and reevaluation of either.

This contrasts with the notion of dynamic synthesis, where both self and rule are redefined in relation to one another through a vital dialectic that reshapes the meanings of the past into a proposed future; the actual is revised in terms of an ideal derived through critical reflection on the way the actual has emerged from the past. The actual is both changed and incorporated, "suppressed and preserved." The culmination and aim of such a process is a stage in which the agent's will is neither pitted as legislator against the actual in the name of an ideal, nor simply subjected to the already institutionalized wills of the other. At such a point, the law is, as Kant insisted, the agent's *own* law, not an external imposition. Here self and other belong to one another and the self is, in the Hegelian sense, "at home" in its world (as this notion is a central theme of chapter 8, I will say no more about it here). Such a harmony is not merely an ideal: there are moments in which our transaction with the world is neither a demanding nor a submitting, but "an experience" (Dewey 1934a, chap. 3) of mutual fit qualified by a sense of aesthetic completeness. Someone whose life is empty of such moments lives a life short of the fully human.

These contrasting relations of self and rule are not fixed stages, but modalities of emphasis within the ongoing process through which self and rules emerge. Their mutually determining relationships are inherently open to change, not because particular selves and rules are only imperfect temporal shadows of timeless exemplars, but because being open and responsive to one another within the process of their mutual emergence is condition of achieving the level of morality. Revision of the meaning of persons in their worlds is a condition of there being persons at all, not a sign that persons have not yet reached an ideal and static form.

Moral Judgment as Both Reflective and Determinant

In a situation where both self and rule are generated and revised within action, no adequate mode of judgment can rest solely on fixed and substantive rules given independently of this action. In Kant's formal terminology, this means that the judgment cannot be simply "determinant." For "judgment in general is the faculty of thinking the particular as contained under the universal," and such a judgment is determinant "if the universal (the rule, principle, or law) is given" and the judgment "subsumes the particular under it" (1790, p. 179). Examples of "given" universals are the categories in science and the categorical imperative in morals. So, for example, for Kant moral judgment has the task of showing how the act enjoined by some particular material maxim can fit under the universal command of reason, itself seen as unchangeable and unchallengeable: "For where the moral law dictates, there is, objectively, no room for free choice as to what one has to do"

(p. 210). If we had (which we never do) a complete understanding of our own motives and of the situation in the world, moral action would not be weighing and choosing and initiating, but *finding* what is eternally true and expressing it in a determinant judgment (p. 267).

Since this sort of subsumption of the particular under the universal is inadequate to the extent that particular acts themselves generate the universal, we must turn to that form of judgment Kant calls "reflective," in which "only the particular is given and the universal has to be found for it" (p. 179). Through this we can take account of the power through which particular actions are legislative as initiating new forms of the universal in terms of which particular actions themselves can be identified and judged—not simply (as they are for Kant) as affirming as their own the eternal dictates of reason.

But we must not go too far in this direction, particularly with the notion that "*only* the particular is given" (my emphasis). Kant himself has taught us that no particular is just "given" as a lump of self-contained meaning, and that something is a particular only insofar as it can be given meaning by being "run though and held together" in a synthesis, that is, grasped by some concept (compare chapter 2's discussion of the ways in which something is a "symptom" only as ordered through the concept of a disease). One cannot simply "have" a morally significant act, then look about to "find" a universal in terms of which it can be judged, for particular acts have moral significance only insofar as they are shaped through an aim at some rule or principle or law. Otherwise, they are merely prudential. So it is as true (paraphrasing Kant) that particulars without universals are (morally) blind as that universals without particulars are empty. This means that the dynamic structure of moral life can be grasped only if we include the processes through which these codeterminate meanings are specified and resolved.

But if acts are creative in seeking a validity that does not rest simply on its reference to a pre-existent universal, then there appears to be neither a limit to the forms of life that act might project, nor any common way of judging them. The similar difficulties that beset aesthetic judgment in Kant's analysis are instructive on this point, for in its capacity as creative the moral judgment now becomes "not a cognitive judgment, and so not logical but aesthetic—which means that it is one whose determining ground cannot be other than subjective" (p. 203). Now some judgments are subjective because their "determining ground" is in the idiosyncratic and contingent aspects of the subject; like those Kant calls the "taste of sense," they are "merely private" and incapable of "making a rightful claim upon the assent of all men" (pp. 52, 54); they report merely that something happens to please one or more people.

Even among aesthetic judgments, however, an alternative to this threatened personal subjectivity can be found in judgments of the beautiful. Like judgments of sense, these refer to feelings of pleasure in the subject, but not to a pleasure whose determining grounds differ from one subject to the other. On the contrary, the experience of beauty is a disinterested delight that reflects the "indefinite, but yet, thanks to the given representation, harmonious activity" of the free play of the imagination and the understanding. These respective roles of imagination and understanding as they are "quickened by their mutual accord" (p. 219) reflect the *Critique of Pure Reason*'s doctrine that the categories can be applied only to a manifold already organized by the imagination. Although in the judgment of the beautiful we do not go on—as we do in cognition—to apply a definite category, the (purported) fact that the form of a given manifold is so *fit* for some category, however indefinite, gives us grounds for demanding that all rational and sentient beings join us in our judgment. For as beings whose experience is organized in time, we share a "common ground" in the categorical structure of the understanding.

The fact that in this apprehension of natural beauty "the Subject feels itself quite at home in its effort to grasp a given form in the imagination" (p. 227) symbolizes (but does not demonstrate a priori) a harmony between who we humans are and the world in which we find ourselves. Exploration of this harmony is part of the *Critique of Judgment*'s task of building a "bridge" (p. 195) between the concept of nature and that of freedom; it is as part of this task that the second half of this *Critique* lays out a principle of teleological judgment. Using this, we can move beyond the minimum order prescribed to experience by the categories to a total systematicity of laws, which have a "unity such as they would have if an understanding (though it be not ours) had supplied them for the benefit of our cognitive faculties" (p.

180). Although Kant himself believed that just such a supreme understanding *had* created the world for our understanding, the principle of judgment must remain merely regulative, that is, a principle that guides us to do science *as if* our understanding could know itself to be "at home in nature" (p. 194) so ordered for it.

A parallel condition limits our judgment of the beautiful, for there "we are suitors for agreement from everyone else, because we are fortified with a ground common to all. Further, we would be able to count on this agreement, provided we were always assured of the correct subsumption of the case under that ground as the rule of approval." But as we can never be assured of this, our judgment "is still only pronounced conditionally" (p. 237): *if* we have eliminated all personal interests and distortions (taken what has long since been called the "aesthetic attitude") and correctly judged that the pleasure we feel is the result of the free play of the imagination and the understanding, *then* we could say that if others were to take the same attitude they would find, in virtue of the common structure of the faculties of the imagination and the understanding, the same delight in the free play of their interaction.

Kant's account of how this all takes place is unarguably the most difficult and arguably the most profound thinking on aesthetics in the whole tradition, so it is a bit unseemly to pass over it without further ado (until chapter 9). But at present I am more concerned with the rich implications of what he says about the form of judgment than about that which is judged. In particular, I focus on his claim that, because we can never be sure that we have correctly identified the source of the pleasure, the judgment of taste "does not *postulate* the agreement of everyone . . . it only *imputes* this agreement to everyone, as an instance of the rule in respect of which it looks for confirmation, not from concepts, but from the concurrence of others" (p. 216). He does not mean by "the concurrence of others" that the judgment predicts the agreement of others, and will fail if they do not agree, since even in the face of their "rude dismissal," our judgment *demands* universal agreement (p. 214). Indeed, it is harder to say what this demand is than what it is not: "It is not a theoretical objective necessity—such as would let us cognize *a priori* that every one *will feel* this delight. . . . Nor yet is it a practical necessity . . . [in which] free agents are supplied with a rule" (pp. 236–37). Nor does the odd status of this always only conditional judgment rest on some notion that the singular experience of beauty in any way creates a new rule or universal: its universal demand rests firmly on the "common ground" that all of us share the timeless structures of the understanding.

It is on this point that I want to take the argument in a different direction: what drove us toward the reflective judgment was precisely the fact that particular acts embodying a new form of universal could not be captured in a determinant judgment. Because of this, the singular act must propose itself as an "exemplar" of a way of organizing experience that "looks for confirmation, not from concepts, but from the concurrence of others" (p. 216). But who are these others who make up the critical community from which our judgment demands agreement? In Kant's analysis of beauty, the communal property that unites them is the shared possession of cognitive faculties. In the present argument, however, the "common ground" must be sought in the common conditions for the emergence and flourishing of the kind of action that is the subject of judgment. Thus it must be sought in the conditions for the correlative emergence of persons whose authority and integrity are inseparable from their treatment of others as persons.

If such judgments were determinant, they would say of the individual act that it was an instance of "treating humanity, in thine own self as well, always as an end, and never as a means only." But since the judgment cannot be analyzed simply as determinant, we must judge the singular act as embodying a way of treating the agent and others as centers of dignity which either gives further significance to an ongoing rule, or initiates new ways of institutionalizing such relationships. So taken, the act is exemplary of the way in which human actions can be organized to make possible the emergence and vital revision of agency. Since this is not simply a determinant judgment that something falls under a fixed rule, it cannot assert that the action definitely does have this power, but must "look for confirmation, not from concepts, but from the concurrence of others" (p. 216).

The point is not that we are waiting on the confirmation of some existent community, and that our judgment stands or falls on what it says, for the critical community called on for concurrence is itself at least partially called into existence by the act. The act embodies and proposes a principle (a "material maxim") that, if generalized and institutionalized, constitutes the organizing principle for just such a community: if others in similar circumstances acted on this principle, the result would be a world of institutionalized relationships that would sustain and foster human dignity.

The fact that this community of concurrence is as yet merely projected and ideal, and may remain so, affects the meaning of the principle to be universalized, for its meaning can be made fully actual only as it is institutionalized and lived in such a community. This is the other side of Plato's claim that no particular fully embodies the Form, for here the point is that any particular embodiment of a rule makes actual only a part of the rules's meaning. Put this in terms of the universalizing implications of act: what counts in the evaluation of an act is the implications which acting on that rule in that circumstance by a person with that history has for acting on the "same" rule in other circumstances by other persons with different histories. Given this, the individual instantiations of the rule are crucial to the meaning of the rule.

It is hardly controversial to argue that the meaning of any rule will vary with the person and the circumstances. Telling the truth to the madman who asks where his knife is (to take Plato's example) gives a different meaning to truth-telling. A rule enjoining individual self-responsibility in which persons are basically responsible only for themselves will have a specific and coherent meaning in a situation where all share the same outlook, and it is possible to succeed with good will and good work; but if the situation becomes one in which others are locked out of the possibility of becoming agents through successful action, then simply acting on the same rule will have quite a different meaning (a comment on "pure" capitalism).

Even if the same person acts on the "same" rule at two different times, the significance of the rule will be affected, for the rule as embodied in act will have a different significance as it is incorporated in the character of the person at different stages of her biography. As Aristotle says, there is a fundamental difference between doing acts "by habit" and choosing them, with knowledge, for their own sakes, and doing so out of a steady state of character (*Nich. Ethics* 1105a30). And there is a further difference between doing the latter for the first time, perhaps in trepidation and against inclination, and doing so with mature confidence. We can imagine repetitive action in static situations where these differences diminish to vanishing. But this is of little importance to the argument, for the moral dilemmas that challenge judgment and define the conditions of responsible authority are precisely those of changing selves and changing circumstances.

In short, it is misleading to think of the rule (the universal) as having a static meaning separate from its embodiment in act, or of an act (the particular) as having moral qualities apart from embodying some rule. Given the interdependence of their emerging meanings, it makes more sense to speak of an "act-rule system" as the primary unit of analysis. Any moral act is structured by an implicit or explicit principle (Kant's material maxim): to be *an* act is to have a form, to aim at an end by some means in some circumstance (the particular without a universal is blind). The act itself affirms this principle, for the conclusion of the practical syllogism is an action that reflects the judgment contained in the syllogism. Put shortly: to act is to judge that some principle ought to inform one's act, and to judge the act—either as agent or observer—is to judge that it embodies the proper principle for the circumstances. It is the making actual of the principle in some circumstance that specifies which of its potential meanings it takes on (the universal without the particular is empty). This means that the judgment (in the act, or of the act) looks for confirmation not in an agreement with a universal already "found" and specified, but in the concurrence of the critical community for which the particular act sets the exemplar.

So interpreted, any judgment of the act-rule system must have *both* determinant and reflective aspects. In one sense, the act does look toward some existent rule for determination: unless the agent "runs through and holds together" the manifold of challenges and threats, desires and aversions, with some focusing principle that generates a situation of moral salience and challenge, no situation will have moral relevance. Here the past has determined the character of the agent for whom some present is problematic. But, in another sense, the rule is generated in the singular act that gives it significance by embodying it in a specific circumstance; this significance can be an old one reaffirmed or a new one proposed. But in either case, the particular is reflected upon and judged proper to be the exemplar for a projected rule. Historically, focus on the determinant aspects has led to theories that see moral argument as seeking justification in the existent rule; focus on the exemplary aspects has seen the vindication of act in future concurrence about its consequences. An adequate analysis must include both the conservative and the revisionist drives.

This revisionist openness to the future does not mean that no specifications can be given on the shape of what will count as moral in the future. If, however, we are to take account of the creative power of human acts, then such specification will have to be formal, rather than substantive. Roughly put, substantive criteria attempt to specify the *constant* value content that any particular act must have if it is to be judged good. In contrast, formal criteria specify procedural rules adjudicating between the *differing* kinds of value that emerge in evolution and history. That these rules are not substantive does not mean that they are "empty" in the sense of useless, for they enable us to take account of human diversity and creativity without falling into a mindless relativism in which, for example, the moral evil of National Socialism is just another "culture." What we need to do this job is a process version of Kant's categorical imperative.

Transition to the Issues of History

The shape of the argument to this point is that the thrust of the self into the world (thesis) is captured in the exemplary aspects of judgment, the thrust of the world onto the self (antithesis) in the determinant aspects. I am building toward an account of how the conditions for their ongoing and vital synthesis provide criteria for evaluating any particular transaction between them, that is, any instance of an act-rule system. Such criteria will have to specify the conditions under which the thrust of the self and the thrust of the other reach an equilibrium within which the "fitness" of each for the other challenges and sustains both. It is because Kant sees something like this mutual harmony as characteristic of the aesthetic moment, and works out a mode of judgment to deal with it, that I have gone to such contortions to see how far a process analysis of moral judgments can be modeled on his schema for judgments of the beautiful.

If we merge these ways of speaking, we can speak of moral approval as the disinterested pleasure resulting from the contemplation of the shape of some rule embodying act. If this pleasure were (as in the usual "moral sense" approach) a simple datum like the pleasures of taste, there would be no grounds for demanding that others agree in this approval. The judgment could do no more than report the occurrence of a peculiar approval, and perhaps urge, in hortatory and imperative language, that others feel the same way (Stevenson 1943). Here we would leave reasoning for rhetoric. But if we take our clue from Kant, we can supply the ground for a moral demand and not just an emotive appeal (which is not to say, of course, that the demand has no emotive appeal).

As the intuitionists have always insisted, the first element in such an analysis can be an immediate sense of approval for some act, whether our own or another's: one simply "sees" the fitness of the action and is pleased by it. But there is also the judgment that this approval is not due to its appeal to any private and prudential interest, but to an awareness that the act is an exemplar of the way in which the moral world ought to be put together. If others were to act in this way in similar circumstances, the resulting community would sustain and facilitate human dignity; hence this singular act harmonizes and resonates with the larger narrative of moral life that it projects, and the immediate and felt approval is a response to the perceived fitness between the shape of the singular act and the shape that any moral world must have if it is to sustain relations of dignity between persons.[1]

I believe it is a matter of plain fact that many lives contain moments of immediate awareness where the agent is at home in the moral world. These moments of closure are of extraordinary importance, not only because of what they can tell us of the conditions that must be sought if they are to occur by choice and not by chance, but because they are the only closure the moral life can have. Our acts seek completion and concurrence in the communities of criticism that they propose, as those communities themselves do in the ongoing processes of vital revision that are history. But history itself is not the exemplar for further instantiations of meaning. Its meaning must remain incomplete, unactualized. Just as an act that does not enter into the living process of interpretation and revision does not "come off," but falls into what Arendt calls those "holes of oblivion," so must history.

This conclusion seems inescapable and disturbing. In terms of individual acts, we surely want to insist that people who act well and die well in the full knowledge that their acts will be unknown and unacknowledged act with moral courage impossible to disparage. But it is not disparagement to recognize the limits of a finitude within which the will can at best impose a partial and temporary order on an indifferent or obdurate world. To accept this finitude as final is neither the sad recognition that some imaginably better state is unattainable, nor a dramatic gesture of defiance in the fact of absurdity. It is not the first, because giving history an ahistoric meaning succeeds only in making the notion of free and legislative acts unintelligible. It is not the latter, with its call for dramatic scorn and defiance, because drama presupposes an audience, and, as Thomas Nagel says, "Such dramatics, even if carried out in private, betray a failure to appreciate the cosmic unimportance of the situation. If *sub species aeternitatis* there is no reason to believe that anything matters, then that does not matter either" (1979, p. 23). The implication is that we should get on with the matters that do count *within* history.

One of the nastiest implications of these limitations to action is the power it puts in the hands of any totalitarian state that can rewrite the past so as to prevent its acts from achieving the meaning they might have called forth (a reflection on the ontological status of freedom of speech). It also lays a burden on the present and future, for the past must not only be preserved and remembered, but given such completions as are possible. It is here that the radical evil of nuclear and other holocaust is most evident: even the present meaning of acts in history is distorted by the threat that our actions may bring about the end of time in which the meaning of action can come to fruition (Stahl 1986). It is here, in this oblivion, that the threat to meaning and motivation is found, not in the absence of some timeless context that assigns history a meaning not its own.

The meanings of history that are indeed its own are in the processes of revision and completion within it: chapter 1 established act as an ordering power, while chapter 2 focused on the ways in which this ordering power could be structured by mutual dignity to instantiate an emergent level of act-world. This chapter's task has been to show that the significance of this level depends on the fact that the organizing power of acts not only imposes a form on the past, but proposes itself as an exemplar to give form to the future. Chapter 4 will explore the status of act in history, where history is the critical revision of the past in terms of this future.

Chapter 4

J. W. Miller and the Midworld of Action

Man has no nature;
he has only a history.
—Ortega y Gasset

Miller's Role in the Discussion

The argument has brought us to the notion of the exemplary act as both embodying and proposing a rule, thus as generating—through the institutionalization of act—the communities within which the meaning of rule becomes actual and seeks authority "from the concurrence of others." The matrix within which this process of concurrence takes place is history, which is not only the product of act, but an inclusive realm within which the meaning of act is completed and judged. To clarify this process, I turn to the categories and distinctions developed by J. W. Miller in a brilliant series of five books, all published posthumously.

Of particular importance for understanding how an act-rule system partially generates its relevant community of criticism is Miller's argument that "history deals with acts. Hence, with purpose. But it deals with purpose as the process that revises it, not as the process that executes it" (1983, p. 32). This process is within the "midworld," which is the evolving context of action midway between—and ontologically prior to—the subjectivity of the self and the objectivity of the other generated within it. These themes carry on the attempt of the first three chapters to restate basic Kantian insights in process terms by showing how action shaped by principle generates a world of self and other within which relations of mutual respect between persons are constitutive rather than contingent.

One reason for devoting this much space to Miller's position is that his analysis is so intense, and his expression so felicitous, that I cannot hope to imitate or parallel it; nor, given the systematic nature of his argument, is it possible to bring his concepts in piecemeal. Further, although Miller is among the most substantive and exciting minds of our times, he is relatively unknown to contemporary discussions. For all these reasons, I give rather more extensive quotes than are usual.

There are, however, some difficulties in expounding Miller's theory of value: first, he did not write one; second, almost everything he says about other matters has implications for value; third, his language is by turns elegantly and darkly Heracleitean, then gently conversational, but never ordinary twentieth-century philosophic prose, much of which sounds as if it were translated (badly) from the German. I begin with a sketch of how his metaphysical and epistemic stances entail a radical theory of value within which "finitude" is the central category, and I will end the chapter by showing how his array of concepts enrich the notion of exemplary judgment.

Miller's Basic Philosophic Stance

Miller sees philosophy as "the science which has no hypotheses," but is "the study of definition by the process of definition . . . [and hence] the study of the universal fact" (1980, p. 41). He quotes with approval Aristotle's contrast between metaphysics and those special sciences that "mark off some particular being—some genus, and inquire into this, but not into being simply nor qua being" (p. 34). From this view of metaphysics as totally inclusive he chides Locke for "assum[ing] special facts, thereby invalidating . . . [his] conclusions" (p. 18). For "Locke, by making mind and object separately definable and interacting entities, put them into the hands of the scientist, for the laws in accordance with which such special entities interact are discoverable only in the usual way" (p. 19). If one begins, as Locke and so many others do, with any of the usual suspects—with mind and body, organism and environment, subject and object—then the fate of Humpty

Dumpty awaits, for "once the split between inner agent and outer fact is made not all the king's horses and all the king's men can patch together the pieces" (p. 72). This means that our most basic notion of what *is*—our fundamental ontological category—must be process, not those elements distinguished in process.

Miller's position parallels Dewey's notion of transaction at the core of this book: in transactions, self and other "do not name items or characteristics of organisms alone, nor do they name items or characteristics of environments alone; in every case they name the *activity* that occurs of *both together*" (Dewey and Bentley 1949, p. 71). As Miller himself puts it: "Object like subject is an omnibus word. If it is to be useful it must be identified through a formal order, as constitution, not as datum or even hypothesis" (1978, p. 109). This suggests that he would have to agree with Dewey's claim that one of the major sources of fallacies in traditional philosophy is "the complete separation of subject and object, (of *what* is experienced from *how* it is experienced)" (1958, p. 32). Nevertheless, he criticizes (even) Dewey for not going far enough, or being consistent enough, on this point:

> What reason is there for basing philosophy on organic response rather than on sunset colors he [Dewey] does not declare. To posit a fact and then to inquire what a fact is, is a logical impossibility. On what grounds, even, can Dewey claim that the responses of organisms are the locus of knowledge? . . . Behind the candid empiricism of Dewey lurks an assumed metaphysics, logic, and the theory of knowledge: facts he must have, cases of knowledge duly accredited as such, a standard for selecting both fact and knowledge, before he can offer a situation which, as real presents the raw material for scientific explanations. (1980, p. 21)

I think Miller has a point here, at least in regard to Dewey's emphasis and language; since Dewey writes from within his own problematic situation against a particular set of opponents, he often does talk as if the organism and environment made up the fundamental elements of process (in order to simplify exposition, I have done the same thing in chapter 1). There is surely a danger that any position that tries to make up for the general neglect of developmental matters will end up taking them uncritically. (I will sort this out in chapter 5 as it bears on the centrality of evolution and development in my own argument.)

In any case, what is relevant for our present purposes is Miller's claim that neither organisms and environments, nor persons and the worlds within which they act, are intelligible apart from the processes within which they mutually define and codetermine one another. We do not first find self and world as independently given entities, then face the problem of bringing them into some nonarbitrary relationship. Failure to see this leaves us suspended between two alternatives, each hostile to an adequate morality: either we are faced with the demand that we should conform ourselves to rules simply "found" in the world, thus surrendering our freedom; or, in the opposite scenario, we find ourselves free agents in a world that is indeed without oppressive rules, but that is also without either guidance or sustenance for our deepest commitments and hopes. Miller argues that an adequate metaphysics dissolves both questions:

> The result of the analysis of the thing paves the way to a theory of value though no word of value has been introduced. It makes man at home in the world, makes it his world. Evolution has its inspiration because it shows the potentialities of slime, and because it makes man of the stuff of nature so that brother fire and sister water are not strange denominations. But if man's mind is alien, then we live externally in a cold environment which is alien to our fundamental value. (1980, p. 155)

Speaking of man as being "at home" in "his world" (the Hegelian echoes are certainly deliberate) is more than a metaphor, for both the world and the selves in it are generated though the same finite and constitutional acts; furthermore, as I will argue in chapter 9, lives and histories can be meaningful in an indifferent universe (question 3) only insofar as they achieve a kind of equipoise in the process through which self and world determine and revise each other.

For Miller, at the core of this process is the fact, so often denied by philosophers, that "finitude is a category," that is, "a factor in the verb 'to be'" (1978, p. 136). Because action is thus constitutional and generative, "we play moral finger exercises with hedonism or obligation, but make no music so long as action is itself disqualified" (p. 93). But to qualify action as constitutional plunges the self and world into the flux of time, with all its risks of chaos and night: for if the law is our own and freely chosen, then we seem to choose at whim, and "Whirl becomes king" in the subjective tumult of conflicting desires and inconstant passions; here moral judgment threatens to become a mere judgment of taste. But if, on the contrary, time and the act are disqualified, then we are condemned to be inconstant shadows without consequence (Plato's problem); here the law we obey is not our own, but an alien other that constrains us from without. How can such a law be both compelling and our own? How can we be, as Saint Paul asks, both under the law and yet free?

The Finite Act as Constitutional

Miller's solution is quite Kantian, a version of what Kant might have said (see chapters 2 and 3 above) if he had studied with Darwin, rejected the mechanical analysis of nature, put (like Cassirer) the categories in history, and thought of the moral self not as banished from time to a noumenal realm in which act is impossible, but as embedded in the time and space structured through act. In essence, Miller works out a process account of how we can be both free agents and yet under the law, "both subjects and legislators in the Kingdom of Ends." The core is the act that articulates finitude, for "in the living moment of assertion resides the true absolute. It describes no *fait accompli,* but an endeavor, and a procedure. When that will knows itself it becomes social, for its freedom can escape subjectivity only as it recognizes its limitation in the will of others" (1978, p. 41). There is "no *fait accompli,* but an endeavor" because both as acting through the "living moment of assertion" and as acting through the actual and institutionalized "will of others," man is unfinished and vulnerable, "a risky, but creative adventure." Here he quotes Ortega's summary judgment that "man has no nature; he has only a history" (p. 105). This is the odd view out within the Western tradition, for as Miller notes regretfully, he did not "find a constitutional incompleteness put forward in any of the types of philosophy. The only discourse that expresses it is history, where there is something less than a *fait accompli* and something more than atomistic unorder" (1982, p. 125). It turns out that the "something more than atomistic unorder" is found in those local institutions that, as embodied act, give the conditions of controlled and responsible action.

So conceived as shot through and through with time, the person has no "environment," no ahistoric locus within which she occurs as a datum: "Let what is considered an act appear as an environed event, and it will disappear into the circumstance in which it is allegedly found. There, its autonomy will be nullified" (Miller 1982, p. 17). This parallels Kant's insight that there is no room in the Newtonian world of mechanistic causes for the autonomous self. But since for Kant this world exhausted the temporal, he had to banish the moral self from time, thus making impossible any account of the moral processes though which growth and creativity were possible. Miller's move is quite the opposite: he so embraces finitude that act, far from being absorbed and nullified in nature as alien, is its very source: "For nature . . . is no datum. It is, perhaps, environment, but even as environment the extension of will. Nature, as order, is pure act. Physics is a study of the general conditions of action" (1978, p. 83).

Where action is generative of the actual, and finitude is a category, the personal pronoun is not a "denotative word" that names a simple entity that is what it is apart from all relationships, but an "organizational word" that "convey[s] extension" (Miller 1982, p. 8). In other words, what it means to be a person emerges out of organized process, and varies within that process with the relationship to others.[1] What Miller says about universals, which are also "organizational words," is illustrative: "The universal is not . . . a class name, like 'apples,' which are not 'pears.' The universal is in the discriminating procedure, not in what is thereupon distinguished" (p. 8). Similarly, for Miller, to be a person is not to be the denotatum of a noun but a present participle that makes self and world actual in time; here "morally" is an adverb characterizing a particular form of self-revising process.

However one says it, the point is that moral theory cannot begin by postulating the existence of a person—the denotatum of some noun or pronoun—who exists and is knowable, but who then must be brought into relationship with other persons or with a moral order. For such relationships, being posterior and accidental to the person's identity, can generate no obligations claiming necessity or authority. If my moral identity is independent and prior to my relationship to my son, or my friend, or my fellow citizens, then I can deny these relationships without threat to my being. Only to the extent that the coherence of my own narrative is shaped by my acts in relation to others am I in the region of necessity and responsibility.

The Role of "Functional Objects"

The point made about persons must also be made about nature: "The order of nature is not an adjective, a property of a *prior* object. Nature and its order are inseparable. Nature *appeared* as order. No one said, '*There* is nature, and upon examination, we find it quite orderly'" (Miller 1983, p. 90). This is equally true of persons and the moral order, for they are "inseparable": persons and a moral order appear and are identified only in a formal order generated through act. The question "Why should I be moral?" disappears if moral action is a condition of being an "I."

Miller calls the formal order within which persons and their other come to be the "midworld," for it is neither self nor other, but locus for the generation of both. What generates them is the "functioning object . . . [which] is that immediacy which embodies the verb—organism, yardstick, clock, balance, number, word" (1978, p. 128). These functioning objects of the midworld generate the formal order of nature through acts such as counting and measuring "that are the actual vehicles of order and are so affirmed" (1983, p. 141). Any act making such an affirmation "becomes constitutional. It projects a world. Any world, any order, is the form of the pure act" (p. 142). The history of philosophy has quite generally and consistently failed to recognize this ontological process: "The great distinctions have had no presentation. Such has been the lack of the great 'categories' from the earliest time to Kant. We have had Space, but no yardstick; Time, but no clock" (p. 59).

Accordingly, I think, ethics must ask, What is the moral equivalent of clocks and yardsticks, of counting and measuring? Miller's answer is clear in outline, but not worked out in detail. He argues that acts toward others (or self) disclose the mutually systematic limits of what it means to be a person in relation to another person (or oneself). These limits, that is, the conditions of control, are embodied in institutions that are the "residue" of action and its embodiment in the midworld. These limits are projected by the act-rule system, and it is in relation to them that its incomplete meanings are given actuality and articulation (cf. the discussion of "act-rule system" in Kantian thinking, "Health and Morality as Levels of Integration," in chapter 2).

The basic implications of his notion of functioning objects can be illustrated with my earlier example of promising, for in that institution I bind my present self—out of respect for both the other and myself—to a specific future. Thus the act affirms both who I am and the shape of the moral time and space through which I act, and within which the meaning of my act can be completed and judged. By defining myself in terms of what I am not yet (myself as fulfilling the promise), I set moral limits to what I can do without contradicting who I am (so far). Since all acts are in principle repeatable (any material maxim can be generalized), any act of promising is potentially an institution: "In all institutions there is a claim to control, to knowledge of truth and of values. . . . They bind the endeavors made in time and finitude to eternal truths and values" (1981, p. 181). This odd phrase, "eternal truths and values," must refer to the absolute status of the agent's commitment, not to some ahistoric status of truth or value (see below, as well as Diefenbeck 1990, pp. 53–58). In this analysis, the institution of promise-keeping is a functional object in the midworld. Where its "binding" is moral and sustains freedom, it is the condition of individuality and immediacy, not their negation: "It is law that makes an immediacy possible. Law does not externally regulate the identifiable datum; it is rather a condition of its discovery. The historic distinction of Kant's ethical proposals rests on his assimilation of personal identity with the social bond" (1981, p. 180).

Kant's (and Miller's) "assimilation of personal identity with the social bond" is the key to answering both the problem of motivation and of freedom. Limiting one's actions through a respect for others does not destroy freedom, for it is only through such control that moral individuality itself is possible: "The will is the force that commits us to the modes of systematic order. That is the reason the will is free. It does not give us what we want, but holds us to those endeavors which permit the emergence of law" (1978, p. 188). What is shaped by law is the flow of ongoing life, not some mysterious moral motive that is a One apart from the Many motives rooted in our inescapable activities, for "a will cannot . . . find its own clarity apart from the circumstances in the face of which it is maintained" (1981, p. 28).

The now familiar problem is to see how moral agency is rooted in and emergent from biological and psychological process. What must be avoided is the "widely accepted psychological reduction . . . [in which this] alleged agency is to be 'explained' by showing that no agency existed, not really" (Miller 1983, p. 39). One mode of reduction is to dissolve everything psychological or distinctly human into the bit parts of physics; another is to reduce mentality to some list of instincts as "atomistic urges" (p. 40). But even at this level, Miller argues for the priority of process. "Instinct," he insists, "refers not to the substance of an urge, but to its status" (p. 41). Any such status will be an equilibrium among the revisions and conflicts of activity. Thus "instinct, spontaneous activity, is the locus of learning. It opens up the environment of the organism. Learning and environment are correlative terms, that is, mutually implicative. Learning does not assume environment, but generates it" (p. 45).

This process of generation is grounded in a "basic and unconscious vitality" (1983, p. 119) giving the organism a "momentum" (1982, p. 14) within which historical and moral action occur as a revision of ongoing process. The ends and urgencies of humans as biological and psychological provide necessary impetus, and material, for further organization. (Cf. the "bioethical" discussion of how the ethical must be grounded in the biological, "Morality as an Ordering Principle," in chapter 2.) This organizational process is historical action, which is the verb to which the moral is an adverbial qualification, for morality occurs only *within* history: "History is not moral: it sets the stage for particular moral systems" (1981, p. 145; see below for the difference between morality as a relatively static moment in process, and as moving and unfinished process). But organisms *are* agents only as they are within history: there is no hypothetical point at which they stand apart from history and decide whether to risk living within it. "Action proceeds from limit, and it arrives at no finality, but only at another defective result. Yet defect, in principle, has no critic. There is no platform beyond limit from which one may snipe at it" (1978, pp. 88–89). Hence criticism is intelligible only from the point of view of the formal conditions necessary for the ongoing revision of action.

This revision of the contingent ends thrown up by the momentum and vitality of living is the essence of action in history. Revision of purposes is essential, for we do not put ourselves in history simply by succeeding or failing in regard to some fixed and finite purpose: "History deals with acts. Hence, with purpose. But it deals with purpose as the process that revises it, not as the process that executes it" (1983, p. 32). Since particular purposes could always be otherwise, they cannot be the source of any such compelling and inescapable modes as history or morality. Thus neither history nor morality can be defined in terms of a set of concrete ends or goals, whether rooted in a static nature or an ahistoric rule: "History avoids finality, establishes finitude, defines the relatively static, emerges from commitment, allies us with evil, and presents the universal as self-revision in terms of the necessary" (1978, p. 92). In the absence of an ahistoric context that furnishes ends in terms of which history can be judged, there is no alternative (chaos aside) but that the criteria of judgment emerge within action as the formal structure of the conditions that make action itself possible.

Such action as can generate its own limits must involve some concrete commitment within time and history: "That morality occurs in limit and only there, that it is not the reaching for a timeless value but for some present, incarnate, and imperfect good, may seem a strange doctrine. But in that way a philosophy of history is extended into the region of ethics" (1978, p. 88). The commitment must be to "some present, incarnate, and imperfect good" because only then is the local act in its irrevocable finitude given generative and constitutional status. This description of the commitment must be in formal terms, since "every specific act emerges from a matrix of commitment, a commitment necessary in principle, but accidental in its content" (1981, p. 33). To say that the commitment is "accidental" does not mean it is arbitrary, and that any choice will equally do. On the contrary, since this act that generates self and world risks the possibility of radical failure, it brings—as a consequence of this possibility of failure—the opportunity for responsible choice: "There is no morality where nothing finite possesses absolute status. Clearly such a claim threatens idolatry, but with no less clarity it is also the condition for discovering idolatry" (1978, p. 94).

What possesses "absolute status" for the agent is her commitment to the "relatively static as the locus of values and the basis of action" (p. 89). Miller explicates this in analogy with Whitehead's notion of the fallacy of simple location: "No location is absolute, but it is just as true that some location must be taken to be absolute in order to make any measurement of relative velocities" (p. 89). Similarly, in politics and in ethics there is no absolute place to stand apart from acts in time, and judge them in advance of action. We begin not from inaction and choices contemplated, but in action where "every moral judgment rests on the base of a current concern, on what one now identifies oneself as doing" (p. 89).

This theme of committed action picks up Aristotle's emphasis on the roles and embeddedness of the person in the polis. Although we post-Darwinians reject Aristotle's notion that the forms of self and society are fixed eternal essences, we ignore at our peril his insistence that moral action has a local place and an animal drive. We do not stand (as Miller says) on some Archimedean platform outside the terror and temptations of time, and then choose, if we are so inclined, the particularities of the human condition. That we act out of, and into, local situations is the compelling point of contemporary emphasis on cultural roots and distinct histories. But what these searches for group identity often lose, or even reject with scorn, is the recognition that all such points are beginnings only, not ends, and that responsible and authoritative action is impossible without intelligent comparison and revision.

In his search for the necessary and the noncontingent, Kant sometimes skimped his account of the nonmoral impetus toward particular ends, and notoriously had trouble getting the agent off dead center and launched into the moral enterprise (his made-for-the-job notion of motivation as "respect" for the moral law seems to have satisfied as few philosophers as Descartes' notion of the pineal gland). Hence his noumenal agent, unlike Miller's embodied one, had no momentum or vitality rooted in the hurly-burly of evolution: "The Kantian morality appears defective on this point. There is no duty nor any rationality until the nonrational and existent moment gives leverage to the moral law. The question 'What ought I to do?' can get no answer, because it makes no sense, until one is already doing something which has for oneself an uncompromising value" (1978, p. 89). What must be shown is how this "uncompromising value" summed up and thrown toward the future by the exemplary aspects of our acts can be both absolute enough to give meaning to our own lives, yet open enough to avoid dogmatism and idolatry. Only then can we avoid a "postmodernism" that can justify tolerance only by surrendering critical judgment.

"Uncompromising" commitment to "some partial aspect of the actual" inevitably risks what Paul Tillich calls the demonic, which "has happened to religion, to nationalism, and even to education. History rides on the vehicles of partial truth, but their demonry is the sole condition of discovering their force" (Miller 1978, p. 90). The demonic aspect is disclosed in the systematic inability of these distorted commitments to sustain the conditions of action itself. Commitment too easily becomes the mindless stupidity of phobic nationalism which decks peasant stupidity out in ritual glory. But the risk is unavoidable: "Life itself is a hazard . . . [which is] absolute, and to that absolute risk there must be an absolute answer" (1981, p. 22).

The necessity of dealing with these particular histories raises questions about how Miller related judgments in history, and of history, to moral judgment itself. Part of the ambiguity in his position reflects the fact that none of the selections in his books are dated or identified as to source (letter? essay? personal reflection?). Such clues would be helpful in sorting out changes in his thought reflecting a different time or context or audience. So in some texts I sense that Miller is his own audience, simply not engaged in an external dialogue that demands definitions or marks the shifting senses between different uses of the same term.

Thus he sometimes speaks of an action that involves values and freedom simply as action in history, yet sometimes he marks off moral action as a subset of action within history. I think we must distinguish two senses of morality here. First, there is morality as a sort of snapshot of a moment of process, which catches the residue of moral action frozen into mores. Here the moral is constituted by institutions and commitments that set an essentially fixed locus for action at some particular time and place: "Morality is a concept of the relatively static. It assumes action within an outlook. It does not apply to the revision of outlooks" (1981, p. 143).

This sense of judgment is "determinant" (see "Moral Judgment as Both Reflective and Determinant," in chapter 3); it reflects one of the alternative relationships between a self and the rule on which it acts: for some acts (for most people in most times) the rule is static and given, and the self's duty is to meet the rule. It is, alas, this concept that has driven much philosophical thinking, and driven it straight into despair at finding such a rule that is neither arbitrary nor stultifying. The alternative is that both selves and rules are caught up in processes shaped by act and aimed at a proposed future; here judgment has an exemplary cast. What I am developing is the argument that we can find our criteria for judgment in the conditions that must prevail if this process of rule-self revision is to be vital and sustaining, that is, characterized by a equipoise in which the self is "at home" in the world.

Miller also contrasts the first and static sense of morality with one involving morality as a process where "revision is the redefinition of the moral, not its detailed execution" (1981, p. 145). Since detailed execution is merely technical pursuit of ends that could have been otherwise, it cannot be the locus of either compulsion or freedom. (Compare the level of practical principles, ranging from the specific and technical to the categorical imperative, discussed in chapter 2.) Because this inclusive sense of the moral as process applies to both moral and immoral actions, it tends to dissolve or blur the differences between morality and history (see the discussion of status and process theories of the state; Miller 1978, pp. 37–41).

Although the moral careers of individuals and groups will involve both phases of relative consolidation and phases of fundamental challenge and change, the second sense is fundamental, for the relatively static *is* and must be judged *as* an extended moment, an equipoise of forces within ongoing processes. These processes, however, are no mere epiphenomenal manifestation in time of what is itself static and timeless, and to which appeal beyond the temporal can be made. As Miller puts it, in one of his favorite phrases, "one does not stand on an Archimedean platform and snipe at the universe or absolve oneself from time in order to estimate the value of its disclosures" (1981, p. 83).

He must have the static sense of morality in mind when he says that "there is no moral judgment upon history" (1981, p. 145), for this would neatly illustrate his argument that "historical events are not to be judged by ahistoric criteria" (p. 83). From *within* history, of course, judgment is unavoidable, systemic, and essential, for history *is* the process within which judgment as the self-criticism of free agents can be maintained: "In sum, history is action, and action is will, and will is both purpose and the revision of purpose, and the revision of purpose is freedom, having no end other than the maintenance of action itself" (p. 35).

The "punishment" for failing to maintain action "is the loss of the capacity to make history, to come to an end of a self-defining task in which lies all compulsion and all universality. These are the reasons why we cannot escape history." It is because of its relationship to universality that this compulsion does not negate agency, but makes it possible: "For the universal is the form of limitation and in all its modes declares the order of critical finitude" (1978, p. 88). Since critical finitude is the way autonomy emerges in time, it is only by making an absolutely risky commitment to some finite enterprise that we institutionalize a world of limits within which self can be articulated. Just as the order of nature emerges out of counting and measuring, so the order of will in its otherness emerges out of the moral acts of which institutions are the embodiment and the condition. Here history is the matrix in which the singular act seeks concurrence in the communities projected by the rules and universals embodied in the act.

This process of seeking is open-ended, for the process of becoming human does not aim at fixed ends: "As a result, one is thrown back upon some intrinsic validity of the will itself, not for a certification of its value by some result external and accidental to it. We can only assert value, we cannot *attain* it or *prove* it" (1978, p. 58). It is at this point that Miller's analysis can be supplemented by the notion of exemplary judgment. His point that we "can only assert value" is essentially a denial that the judgment can be *determinant,* which it would be only if "the universal (the rule, principle, or law) is given" and the judgment "subsumes the particular under it" (1790, p. 18). Although Kant saw moral judgment as determinant, Miller cannot, given his claim that ends themselves evolve within the process of will as it is affirmed in the institutions of the midworld. The flux of this situation can be grasped only it includes a "reflexive" judgment where "only the particular is given and the universal has to be found for it" (1790, p. 18).

The challenge to Miller's position (and mine) is that if the will has no ahistoric environment of universal ends, then it threatens to become merely capricious, even demonic and idolatrous: "The forces it lets loose seem mad, arbitrary, and uncontrollable" (1978, p. 142). This is quite like Kant's problem with the judgment of *taste,* which is "not a cognitive judgement, and so not logical, but is aesthetic—which means that it is one whose determining ground cannot be other than subjective" (1790, pp. 41–42). This is *precisely* the problem Miller embraces—and solves—with his emphasis on finitude as categorical, where the subject is "a factor in the verb 'to be'" (1978, p. 142).

We can capture this originary and projective power of act in the notion of exemplary judgment in which we are "suitors for the agreement from every one else, because we are fortified with a ground common to all" (1790, p. 82). For Kant, that common ground was in the conditions of judgment in general (the fixed categories), but for Miller it must be in the conditions of sustaining free action. Our judgment and our commitment is to some "present, incarnate, and imperfect good" that we take as exemplary of the ways in which human action can be organized and institutionalized to sustain the emergence of agency. Such a judgment, as Kant says, "does not *postulate* the agreement of everyone. . . . It only *imputes* this agreement to everyone, as an instance of the rule in respect of which it looks for confirmation, not from concepts, but from the concurrence of others" (1790, p. 23). For Miller, this concurrence of others is the judgment of history, within history: "All history displays this equation between selfhood and actuality. This is the stage on which all action occurs. . . . If we can find the actuality which is for us self-defining, and hence absolute, we shall know what to do. Otherwise we may well suffer disintegration and defeat" (1978, p. 87).

Transition Back to the Original Questions

My argument sounds like the usual pragmatist claim that evokes the criticism that pragmatism always puts off confirmation and validation to some further event: the check for value is always in the mail. In a sense, this is right. If the meaning of act is exemplary, and calls on the future for completion, then the end of history will leave human meaning incomplete. But this does not mean that within history there are no acts and lives that have all the meaning that finitude permits.

The construction of the midworld within which self and its limiting other come to be involves continual revision, both of what the past comes to mean as it is incorporated in the present, and of what the present strives to mean by projecting a particular future. There can be moments within which the projected future and the incorporated past come together in "an experience" in which human meaning is exemplified. Although it is not "completed" in the sense that no future can expand its meaning, it is "complete" in displaying a harmony between what the self demands of the world and what the world demands of the self. To explore how this kind of ongoing completeness is possible within history in an indifferent universe is one of the main tasks of the present argument. Since the shape of this task is set by the three original questions, we must turn back to them to see what dimensions they take once we incorporate into them the results of our arguments so far.

Chapter 5

The Constitutional Status of the Three Original Questions

> The Spirit is never at rest, but is always engaged in ever progressive motion, in giving itself new form.
>
> —G. W. F. Hegel

The Status of the Original Questions

I began this book with three questions:

1. Is the belief that people can make responsible and autonomous choices about their lives compatible with what we know about the evolutionary and developmental processes out of which people emerge?
2. How can the choices made within biological and cultural history be nonarbitrary and compelling if there are no ahistorical norms against which they can be judged?
3. What sort of meaning can individual lives and choices have if the history within which they come to be is simply an accidental episode in an indifferent universe?

I assumed that these questions were, though not particularly well defined, at least a familiar way of expressing concerns shared by many contemporaries—by no means all of them philosophers. Such a rough-edged procedure was unavoidable as a formulation of the problematic situation that began the inquiry. But now we must ask just why *these* questions, when others would highlight alternative issues and precipitate a different book? After all, there are many thoughtful people whose concern is to *purge* ethics of every taint of science, not make them compatible. What strikes them as salient is not what strikes me.

I obviously hope that the same questions are urgent for my readers as myself (otherwise, no readers). But what is the status of our shared interests? Are we simply people who happen to be puzzled about the implications of rationalist Western philosophy, respectful of the developmental sciences, nonreligious, tolerant of semi-Kantian jargon, and curious about odd metaquestions? The more I am convinced that these are the right concerns to start off with, the more I am nervous about my own confidence, for beginnings are all too likely to determine ends.

We must look both at the urgency of the original questions and at the ways in which their exploration leads to their revision. I believe that while their particular form reflects a rather local situation, they point to underlying and unavoidable considerations that must be dealt with by anyone who takes time and process seriously—and also by anyone who doesn't, since part of the meaning of taking self and other as unemergent, fixed entities is in what this denies. Both alternatives raise the issue of whether "taking process seriously" doesn't itself beg substantive questions.

The Questions as Representing the Dialectic of Process

It would be a serious matter if the three questions were urgent only for those who happen to share the cultural and biographical experiences out of which they arise. For what does not compel our concern can be ignored as irrelevant. In Miller's words: "Philosophical . . . authority and compulsion must wait on each person's acknowledgement of his own identity with the problems" (1978, p. 72). If, however, there are (as I shall argue) ends that we share simply as being human, then questions concerning them will have the requisite "authority and compulsion."

Traditionally, such ends have been proposed either as commanded by God or as built into a fixed human nature, but neither option is open to the present argument. For since, as Ortega says, "man has no nature

. . . only a history" (1941, p. 217), there is no fixed human nature in which common ends might be grounded. Hence what we share cannot be common ends fixed by a common nature, but the common necessities of revising and ordering whatever ends are thrown up by the transactions out of which both self and world are emerging.

Since each of the questions points to elements presupposed by this process of revision and ordering, they are not accidental, but (in one form or another) unavoidable. Their particular formulation is, of course, far from unavoidable, and reflects a local situation in which a respect for the power of evolutionary and developmental sciences is seen in tension with more traditional concerns. But, taken jointly and in relationship to each other, they reflect the inescapable dialectical elements of any human experiencing (of any midworld, in Miller's terms). Thus question 1, the thesis, focuses on the nature of the agent who acts into the world; question 2, the antithesis, deals with the shape of the world within which the agent acts and experiences; question 3, the synthesis, concerns the ways in which the nature of the agent and the world affect each other.

In Miller's terms, these are "constitutional" issues, unavoidable because they deal not with some particular purpose but with the systematic means for revising all purposes. Revising purposes does not go on simply because of the contingent fact that selves exist and have a difficult time choosing among their many ends. On the contrary, there *are* no selves apart from the ordering of world and self that comes about through the pursuing and revising of ends (see chapter 2). Hence self and world have no being apart from these activities, and are thus *constituted* through them, and questions that deal with the conditions for the generation of self and world are "constitutional."

Miller gives a quirky but rich example of the shift in significance from simply having ends as a contingent and unexamined matter of fact to revising them. Only the latter puts one in history, for this "deals with purpose as the process that revises it, not as the process that executes it" (1981, p. 32). Suppose that I wish to play tennis and choose flat-soled shoes as a means to this unquestioned end. My choice concerns means only, and success or failure involves "technical" matters, imperatives of skill. The same is true of choosing some means to win a battle or get new territory. At this level there

> is no way of judging the value of playing tennis, or of winning a war. One simply begins by declaring tennis or war. In so far as the tennis and the war reflect no larger environment to which they are adjustments and expressions, they are not values, and hence cannot be said to be purposes. They cannot be said to be *acts*. No *person* is merely a tennis player or a soldier. A person plays tennis in the interest of a more inclusive program. Were this not the case, one would be back to mechanism or to catastrophe. (1981, p. 32)

The phrase "declaring war or tennis" is just right to catch the sense of whim that would characterize such an event (not yet an act) not embedded in past or future. No such merely momentary twitching would mark out an individual organized and oriented enough to "take an interest" in something guided by "practical principles," that is, "propositions which contain a general determination of the will, having under it several practical rules" (Kant 1788, p. 18; see chapter 2). This could only happen if there were some center of action and effort to "run through and hold together" the items that fit under the several rules, which would themselves be united under one general determination. Without this, only "mechanism or catastrophe."

The tricky notion in this passage is that the particular is pursued "in the interest of some more inclusive program," for this opens up the question of what the *most* inclusive program ought to be. I think the general answer is something like "the most inclusive program is becoming and staying human," or, as Miller would say, being in history. Essentially, the inclusive program must be formally defined in terms of the qualities of integration that will sustain the emergence and revision of meaning as a distinctive human achievement (of this, more later).

Our present concern is with the intermediate level, where organizing and coherent action is less than fully inclusive: "The game of tennis is no absolute. It suits one's situation. It is played because it suits" (Miller 1981, p. 33). That it does so is not a matter of wholly conscious choice. After all, one inherits a language and a social order and a set of roles, as well as a sequence of DNA, and one simply finds oneself in a situation in which tennis, not cricket, is played: "Every specific act emerges from a matrix of commitment, a commitment necessary in principle, but accidental in its content" (p. 33). These commitments are the "mores" part of morals; as such, they define the limits of a world within which impulse and momentum can come to be act. They reflect the determinant aspects of judgment in which both self and other are, for the moment at least, constant and unchallenged.

This still does not fully capture Miller's point, for he not only makes note of the fact that we inherit a mixed bag of institutions, but says that this is in principle unavoidable. It is not just a (mostly unhappy) fact that agents act in contexts that are not the result of their own acts, but that human actions *must* do this if there is to be responsible human action at all: this is the condition of being embodied, and without it there is no vital encounter out of whose dynamics the self as ordered can arise.

The point of saying that some matrix is unavoidable in principle is sharpened by the fact that each context of action is "accidental" in its content. It is here that many a skeptic has taken a triumphant stand, insisting that the fact that everyone (even Aristotle) speaks from some particular perspective in time and space limits the relevance of what they say to those who happen to share the starting point. Questions within this problematic situation will have no urgency for those who do not share it—as we in the twentieth century do not share Aristotle's. Nor do skeptics or religious thinkers share the specific form of the questions beginning this book.

The answer is that the human situation itself is inescapably urgent: the pressures of finitude make necessary the revision of ends without which neither selves nor worlds can come to be. Miller's next point bears on this, when he notes that tennis falls under a "program under way"—something like going to college in a specific culture. But whatever the "particular view of things that leads to tennis" is, it is no absolute. Both it and its components are

> necessary in principle and accidental in their peculiarity. . . . For that reason they are all unstable. Their contingency (or accidental aspect) bespeaks their restless demand for enlargement. To become conscious of the accidental is to see its limits; and to see limits is to apprehend the next factor, the environment of a limit.

> So one sees that one goes to college because of the times and the place, but one sees that there are other times and places, and thus going to college becomes one's conscious will, or else wins rejection as something unsuitable. (1981, p. 33)

This means that the accidental aspect of the context of commitment does not condemn it as arbitrary and apart from criticism, but is, on the contrary, precisely what is needed to generate that critical community within which responsibility and autonomy are articulated. It is just this recognition of finitude and accident that breaks the hold of that which is given contingently and accepted uncritically. In Plato's Myth of the Cave, this is the "happy chance" that makes it possible to break the chains of habit and see that the shadows on the wall are "all unstable," the result of accidental perspective whose "contingency (or accidental aspect) bespeaks their restless demand for enlargement." For Miller, of course, there is no escape from the Cave of finitude into the unchanging and reassuring light of the Sun, so "enlargement" comes not from the transcendence of all perspective and restless momentum, but in continual and responsible revision of purposes where the shape of some perspective "becomes one's conscious will, or else wins rejection as something unsuitable."

Miller's description of a situation in which we seek more and more inclusive ends, and ask increasingly general questions about values, threatens us with the (traditionally unanswerable) question about the nature of the most inclusive end. But he is not seeking the unity of a common and inclusive end (rooted, perhaps, in a common human nature), but a unity imposed by the process of revision we share in our pursuit of the uncommon ends in terms of which we work out our different modes of being human:

> Whether one deals with the acts of individuals or of groups, this revision of the basis of action is necessary, for it defines action. Without it, there is no purpose. But what governs this revision? Merely the necessity of will, its search for itself, for its clarity in detail, and for its own being as well. History is will. It is the career of will. History is the progressive clarification of will. It has no ulterior goal. (1981, p. 34)

This claim is central for any narrative of how distinctly human animals (question 1) can make a meaning for themselves within a history that "has no ulterior goal" (question 3) and does not reduce choices within history to "mechanism or to catastrophe" (question 2). The specific dilemmas driving those three questions do not need to be universally shared for them to have a general and necessary relevance. For while each of them reflects the blockage of momentum in some specific matrix of commitment that is accidental rather than absolute, they jointly reflect the unavoidable questions raised by the process of revision in any context whatever. Thus the specific questions are local forms of general questions that are unavoidable for understanding any process within which critical revision—and hence humanity—is possible.

It might be objected that this end of making sense of the human condition is itself accidental and avoidable. But the critical revision involved in "making sense" of emerging humanity defines the human project; it is not just accidental and contingent. It is not that there are first human beings, then they happen to raise questions about the meaning of their lives and acts, about structure and significance of the world, and about how the two fit together. On the contrary, raising such questions is inseparable from the emergence of distinctively human animals: it is just these focusing activities that are essential if the requisite levels of organizational stability are to come into being. So by "inseparable" I do not mean that they always happen to go together, but that the questioning is an essential part of the process that *constitutes* what it means to be this sort of animal. There is a Cartesian twist here, for questioning the necessity of questions about human meaning is already an instance of the unavoidability of questions about human meaning.

The terms Dewey uses to describe this process are "doing," "undergoing," and "experience." In chapter 4 I touched on Miller's claim that Dewey reads these terms too much at the level of the biological thrust of the organism into the natural environment, and not—as Dewey's own theoretical position would entail—at a level of ontological egalitarianism that leaves open the possibilities of different kinds of entities on either pole, and of form-imposing activity being due to either pole, not just to the biological "act" of an organism adapting in nature.

To fix on the kinds and relationships of entities salient within a particular problematic situation, and consider them to be, without qualification, the kinds of entities central for *all* experience, would be to fall into what Dewey calls the fallacy of selective emphasis (1958, p. 27).

Philosophers do this with dismaying regularity. From the fact that entities at the biological (or microphysical, or social, or whatever) level are what must be given selective emphasis for understanding one range of problems, they conclude that only these entities are "real," and hence that moral and aesthetic entities are derivative and epiphenomenal (see chapter 1). The equally convincing (and misleading) counterclaim of those who have cut their categories on problems of the "human spirit" is that only intentional entities are fundamental, and that the biological is ultimately derivative, a mere aspect of something more fundamental. Such disputes keep many good folks employed and the journals full, but do little else.

But the fallacy is seductive, especially for any argument that puts major emphasis on one form of knowing, as the present inquiry does on the developmental sciences. The temptation is to take the categories appropriate to evolution and child development as descriptive of the most "basic" kind of existence, then to fit (cutting off the protruding edges) all other human activity into this mold. This would be a clear instance of letting the original shape of the questions distort procedures of investigation in question-begging ways.

Consequently, it is essential to emphasize that the "generation" of self and other is a methodological and ultimately metaphysical concept; it makes no claim that the developmental and biology story has a privileged access to the real. Such formal terms as "doing," "undergoing," and "experience" refer to functions that can be expressed at a multiplicity of levels, not simply biological and developmental. That they are organizational and not denotational terms (Miller 1982, p. 8) is reflected in the fact that there are no pure and isolatable instances of them: there is no doing (of a person or a nation) that does not undergo the limitations of its relation to some other; there is no undergoing that is not the undergoing *of* some other's doing, no experience that is not emergent from and qualifying of specific interpenetrations of doing and undergoing. As Dewey says,

> "experience" is what James called a double-barrelled word . . . [which] includes *what* men do and suffer, *what* they strive for, love, believe and endure, and also *how* men act and are acted upon, the ways in which they do and suffer, desire and enjoy, see and believe, imagine—in short, processes of *experiencing*. . . . It is "double-barrelled" in that it recognizes in its primary integrity no division between act and material, subject and object, but contains them both in an unanalyzed totality. "Thing" and "thought" . . . are single-barrelled; they refer to products discriminated by reflection out of primary experience. (Dewey 1958, p. 8)

Although it is appropriate to limit the term "experience" to conscious human experiences when we are talking about the subject matter of the three questions, the dialectic of process obviously extends more widely than this. As the argument for the more inclusive senses of doing, undergoing, and synthesis develops, it will become clear that part of the justification for using these categories at the human level is that this allows us to assimilate the human experience into a broader sense of natural process. Ultimately, the full justification for "taking process seriously," whether at the level of human experience or in the broader metaphysical sense, rests on the fact that doing so enables us to resolve the particular problematic situations that drive us to thought —and to do so in a way that is also capable of resolving not just this local way of being puzzled and challenged, but any legitimate way as well. In this sense, the test of the soundness of the present analysis is quite pragmatic: does it break down blockages in our thinking so that we can better use our knowledge about ourselves and our world to revise our purposes in vital and sustaining ways?

Reformulations of Question 1

"Is the belief that people can make responsible and autonomous choices about their lives compatible with what we know about the evolutionary and developmental processes out of which people emerge?"

Even if the first question does not beg the general methodological issue about the status of the developmental sciences, it does make assumptions that must be made explicit and revised. For example, it assumes that the possibility of responsible human action is threatened by the fact that humans emerge out of ontogeny and phylogeny. But every argument we have looked at shows that embeddedness in the process of articulating self and world is the *condition* of responsible human action, not a threat to its existence. Thus the basic evolutionary and developmental story of how distinctive forms arise within processes includes that of the ethical as itself an emergent, irreducible, and methodologically unproblematic level of activity: the distinctive modes of human functioning as shaped by ethical rules and institutions are an "achievement" of a level of integrated and systematic functioning that has distinctive powers and properties not reducible to the elements out of which they emerge. In their functioning as systematic entities, these persons must be understood as embedded in the appropriate level of space and time defined by the institutional residues of acts, and must be analyzed in accordance with the formal principle that self and other are codetermined in process.

Although this principle of codetermination emerged out of an inquiry formulated in terms that saw animality and embeddedness in time as threats to the distinctive power and autonomy of ethical action, it is not limited by the shape initially given to the problem. For as the argument has developed, the question has moved from being a local puzzle about the relation of biology to being good, and become a question about the conditions of being human. And these questions, regardless of our answers, are constitutional and unavoidable: no matter where we begin, we must ask whether the self is capable of action in the world (question 1). Similarly, we cannot avoid (question 2) asking whether the world is such that it can sustain and support the intentions of this action, nor (question 3) asking about the implications of each of these for the other. At this level, we are not inquiring about some particular purpose invited or stymied by the shape of a particular world, but about the ways in which all and any human purposes can be pursued and criticized in the world as it is. This involves a shift of emphasis away from actions that are moral or immoral in specific departmental senses, that is, moral as opposed to cognitive or aesthetic, to actions that are generative of human selves in a human world that includes aesthetic and cognitive and moral dimensions. In Miller's terms, we have been talking more and more about the conditions of being in history, and less about the specific conditions of being in history as morally good persons. This will become more explicit in the consideration of question 3, which concerns the conditions for the synthesis of issues raised in question 1 as a thesis and question 2 as an antithesis. These dialectical relations will assume increasing importance as the argument progresses: in chapter 8, I tie them to parallel distinctions between three forms of truth (truth of statement, thing, and spirit) in Albert Hofstadter's metaphysical dialectic; in chapter 9, I argue for the methodological centrality of Socrates' question, "How should one live?"

Reformulations of Question 2

"How can the choices made within biological and cultural history be nonarbitrary and compelling if there are no ahistoric norms against which they can be judged?"

The basic litany here is the same as in the first question: what is presented as a threat turns out to be the core of the answer once the problem is reformulated in process terms. Although the original formulation reflects a particular problematic situation, the issues raised are systematic and constitutional.

The misleading assumption is that norms can be nonarbitrary and compelling only because their "ahistoric" status gives them an authority that does not depend on the actions of any subject in history. From this perspective, no meaning that agents give to their lives (question 1) will be coherent and sustainable in the face of disaster and dissolution (question 3) unless it is ratified—essentially duplicated—in a realm totally independent of the agent. Action is disqualified from being coherent and choiceworthy precisely to the extent that its choices and coherence reflect the dynamics of action itself. Thus, oddly enough, although humans are of supreme worth largely because of the kinds of actions they do, these very actions are seen as powerless to generate the criteria that define this worth.

An examination of the dialectics of doing and undergoing, however, shows that formal and nonarbitrary criteria for evaluating action can be derived from the conditions necessary for selves to emerge and function in relations of mutual respect and dignity. These relations give the conditions for critical authority in *all* realms of human action, not simply the moral. The focus is on the overall process within which subject and other come to be, and within which they have a significance only in relation to one another: just as autonomy is always *within* some context, so responsibility is always responsiveness *to* a specific set of demands integral to the emerging self in that meeting these demands is a condition of the self's own integrity and authority.

One reason for accepting this account of the generation of norms within history is the lack of any defensible account of how they can be found outside of it. But even if one accepted—whether on argument, faith, or "intuition"—that we can find such norms and know them as they are "in themselves," their compelling power would be an autonomy-stifling external compulsion, a mere threat of punishment or promise of reward. For if norms are objective in this sense of owing nothing to the conditions of the choices they purport to guide, then their relationship to the will of the agent is external and contingent. They are what they are quite apart from any action or intention of any (at least finite) agent.

Where the relations between norms and agents are external in this way, then so are those between different agents: if my identity is independent of any relation to other people, then in principle I can always do without them, however much they may at times contribute to my felicity. So I might feel sympathy with someone and act with care and kindness toward her; our interests might converge so that the pursuit of my good is compatible with—or even requires—her's. But if neither sympathy nor convenient overlap happen to happen, then my concern for her can only be imposed by some source other than the self. With this, action is compelled; absent this, action and judgment are indeed arbitrary.

I have argued that Kant correctly saw that a source of law other than the self was compatible with freedom only if it was at the same time identical with some aspect of the self. Thus chapter 1 argued that treating others as centers of dignity gives rise to a level of organization in which persons emerge out of the merely animal and prudential. Chapter 2 dealt with ways in which other-regarding actions generate both moral self and moral other. Chapter 3 looked at ways ethical judgment in its exemplary aspects reflects the incompleteness of the emerging self embedded in process. Chapter 4 related these claims to Miller's analysis of the ontological status of history, and to the notion of moral acts and institutions as the "clocks and yardsticks" through which the midworld of action is articulated.

These chapters were responses to the three questions, and the present chapter has argued that these questions are themselves part of the process of articulating the moral self through the revision and criticism of human meaning. The dialectical nature of this process is displayed in the relation between the demand of the autonomous self that the world conform to the "ought" laid down by the self, and the ways in which the world apart from the self answers or mocks these demands. These issues are explicitly brought together in question 3.

Reformulations of Question 3

"What sort of meaning can individual lives and choices have if the history within which they come to be is simply an accidental episode in an indifferent universe?"

This assumes—quite reasonably—that the significance of individual choices must be understood in terms of their role in the emerging narrative of a person's life. Meaning generating choice cannot occur in isolation: it must engage the past and propose the future—hence the originary character of act and the exemplary character of judgment.

The further assumption is that just as individual acts are qualified by their incorporation into autobiographies, so are these latter by their incorporation into history as a whole. I have no quarrel with this, but only with the notion that this inclusive process has a significance itself not generated out of its constituents. For thinkers and plain people alike have feared that if the meaning of history has no other warrant than what is proposed within it, then it will be a "tale told by an idiot." In this view, we can live with authority and responsibility only if our ideals reflect a meaning guaranteed by God, or the inherent moral structure of the universe, or even by some fixed purpose within history (à la Marx), just so long as it is not proposed or revised by what happens in history.

I do not reject this for its opposite, that is, the counterclaim that the meaningfulness of human life is not at all a discovery, but is a pure projection of autonomous spirit. To do so would reflect a cancerous and destructive proliferation of the claims of the self (raised in question 1) to be autonomous and responsible for its own will and decision. The will of self as law-giver and world-creator would be unresponsive to any limitations set (question 2) by the structure and norms of an other. Self would be arbitrary in the imposition of its will, lacking not only external compulsion, but any compelling reason to consult either the will of another or a norm not subject to its own whim. Meaning and coherence, however, can be achieved only if there is a limiting other that frustrates or sustains the agent's actions, for action is not fantasy, but finitude. Hence the emergent meaning is a synthesis of the thrust of the self into the not-self as challenge and resource.

What is needed to make sense of human action is an account that denies both the unrestrained claims of the self to autonomy and the unrestrained power of the other-than-self to compel, and yet preserves some element of both in a synthesis. The self that thrusts into the world has only a limited power to impose order upon it. Neither the self's own character nor that of the midworld it reflects and structures is made *ex nihilo,* but out of the limiting potentialities given shape through action. It is for these limited contributions that we can meaningfully be blamed or praised.

Any autonomy is limited by compulsion from without, the counterthrust of the other that the self undergoes, and that makes it impossible for the self's act to be fully its own. The dialectical pair here is autonomy **: :** compulsion, and the required synthesis will be a compulsion that does not destroy autonomy. This is found in the institutions embodying the actions of the self (they *are* the self in its otherness), and that articulate the ways in which its own dignity and autonomy are structured through the recognition of the autonomy and dignity of others. The realm of such institutions is history.

Consider again responsibility, that other aspect of the self as autonomous. Responsibility is denied if there is no other *to whom* the self is responsible. With no norm to set limits to the self-aggrandizing will, all choice is arbitrary. The dialectic pair is responsibility : : objectivity, and the required synthesis is a limit that is itself responsive, an other that enables the very act it conditions. This, too, is found in the institutions that are the self in its otherness, and that articulate the ways in which its own dignity and autonomy are structured through the recognition of the dignity and autonomy of others. The realm of such institutions is history.

The radical opposition between compulsion and autonomy and between the arbitrary will and responsibility are overcome, yet preserved, in the notion that the structure of the other is not that of the other as it is in itself, but of the other as it emerges from those transactions through which both self and other are codetermined. Ideally, action would be limited only by those conditions integral to its own emergence: here the other that gives the self the limits that elicit and define its existence would be the self's *own* other, not something alien and external.

But we do not live in the ideal, although we have moments in which doing and undergoing are fit with each other, thrust and encounter perfectly matched. Such moments confirm and celebrate the fact that we can be "at home" in our world; here what the other demands of us we offer freely, and what we need of the world is freely given. It is this harmony that Kant explores in the principle of judgment, and that Dewey describes in *Art as Experience,* where he argues there that the qualities of wholeness and completeness that paradigmatically characterize aesthetic experience are the key to understanding these more inclusive experiences. Because the aesthetic quality of "an experience" indicates a *way* of experiencing, rather than the experience of a particular subject matter, it is possible that the life of action or of knowing, or ultimately the life of the self in all its complexities, can reach the kind of unity often more intensely reached in art.

The point is not that our lives have meaning only in such moments, or would have meaning only if the whole of our lives had such an adverbial quality. But such moments do indicate that (in crude terms) we are "doing something right." In the bioethics language of chapter 2, they are a phenomenological sign of the full functioning of the "healthy" self in a supportive world. But although they occur in many lives, they vary much in degree and frequency; too many people, too much of the time, can lose the tension of disequilibrium with the world only through drugs or denial, not by bringing it into a vital synthesis in action. And for most of us, living is not made up of moments of confirmation and celebration, but of routine and largely unexamined activity, pleasant and fulfilling if we are prudent or lucky, deadening or deadly if we are neither. Having a meaning for our lives in this minimal way is just having some story, however sad or inarticulate, that holds events and choices together as our own, even in a harsh and hostile world. But even this minimal achievement involves the notion of synthesis, since the articulation of even the most sorry and pathological self presupposes *some* congruence between what the self demands and the world answers, and what the world demands and the self has to give. Even merely physical survival presupposes a survivable niche.

Transition to Issues of Methodology

These points remind us how easy it is to fall into the habit of talking about "synthesis" simply in honorific terms, forgetting that some level of synthesis is involved in every organic activity. Even a psychotic hallucination, a projected mental content totally unresponsive to the shape of the world or the significance of sensory input, depends on whatever context makes it possible; even the most brutally compulsive subjugation of the organism to an external power is the subjugation of just *that* organism. Obviously, not all these modes of activities are human activities, let alone sustaining human modes of action exemplifying the meanings that hold together a fulfilling life. What is involved here is not yet clear, except in outline, and must wait until chapter 8's discussion of "consummatory" experience to be filled in. But it *is* already clear that the significance of such experience must be sought in the dynamic structure of synthesis, not in the nature of self or other in isolation.

It is also clear that any description of the structure of synthesis within which the self is at home in its world must be formal and abstract enough to take account of the fact that this process of self and other coming to be in relation to one another is open ended; not only this, but these relationships must be subject to continual critical revision if their mutual equilibrium is to be vital and distinctly human—to be *history,* in Miller's sense. If, however, we can get a hold on the relations between self and other that are necessary to sustain the vital equilibrium within which living is coherent and meaningful, then the formal principle defining such relationships can serve as a criterion for judging different ways of living. This would extend the argument of chapters 2 and 3 that relationships involving the recognition of mutual dignity are the condition for the emergence of persons as moral agents.

Since this process of human emergence is rooted in the biological and psychological, I turn to an examination of the limits and structures of synthesis in these areas. The results will guide our extension of the analysis to those processes that achieve modes of knowing, acting, and appreciating that define the inclusive sense of being human.

Metaphor; example:
sailing a boat in a race
efficiently
dynamically } responsive to wind, water,
alertly geography, crews characteristics,
time etc

Chapter 6

Self and the Focus of Significance

> What I wish to point out is that the scientific method is, after all, only the evolutionary process grown self-conscious.
>
> —George Herbert Mead

The Status of Inquiry in the Developmental Sciences

The last chapter argued that the three questions had a "constitutional" status derived from the fact that they formulated unavoidable elements in the dialectic of process.[1] Thus the fact that they were stated in the language of evolution and development reflected the concerns of the local problematic situation, not an assumption that the categories of the developmental sciences are related to the world in such a direct and privileged way that all other inquiries have to be translated into them as a condition of being taken seriously. Insisting on such a reduction would commit the "fallacy of selective emphasis" (Dewey 1958, p. 28) by assuming that a set of categories selected as primary and salient for one inquiry were equally fundamental in all.

Although the power and sweep of the scientific categories involved make this sort of reductionism a continuing temptation (hence sociobiology), our concern here is not with the effects of taking the developmental sciences as ontologically prior, but with what it would mean *within* these sciences to acknowledge that they have no such privileged access to the real, but are—like all inquiries—the articulation of critical perspectives within a changing and emerging problematic situation. When we have done this, then we can incorporate the rich insights these sciences have into our broader normative inquiries without reducing the biographical to the biological.

I have neither the competence nor the space to describe what the developmental sciences would look like if they systematically saw themselves as operating within this process framework. Fortunately, however, we can get at a number of the issues involved by examining how the notion of "normality" functions in various inquiries. For if an inquiry neglects the roles of focus and context in determining the structure and makeup of its field, this is immediately reflected in static and uncritical notions of what it means to be a normal entity or behavior in that field—or in what makes up a normal field. At one level, of course, this is inevitable, for contexts *do* remain (relatively) constant, and for simplicity's sake we talk about matters like health or adaptation (or morally admirable characteristics) in and by reference to some taken-for-granted context. It is all too easy, though, to slide from this to taking the properties reasonably attributed to some entity in the normal context as properties of the entity *simpliciter*. It is equally easy, and potentially as disastrous, to assume that the significance that a context has for some "normal" entity is the same as its significance for some other entity, or the same entity at a different stage. A trite example: an evolutionary modification may be adaptive in one context, a death sentence in another, just as what is an environmental threat to one organism is an opportunity for another.

Criticizing the ways in which static senses of normality wreak methodological havoc will make clear why they must be replaced by the notion of "locus of significance." This concept organizes inquiry into the processes through which entity and context come to be in relation to one another, not just in the developmental sciences, but quite generally. Ultimately, I will argue that the very meaning of what it is to be a particular kind of subject or object is specified in terms of their joint and mutually defining contributions to some process taken as the locus of significance.

At least three assumptions adversely affect how the concept of "normality" is used in developmental theories and research: (1) organism and environment can be treated as separate entities; (2) similar outcomes are the result of similar processes; (3) science is value free. Since these assumptions are pivotal to what we mean by "normality," the confusion they generate infects its usage as an authoritative standard; an index of health, physical or psychological; an index of acceptable physiological function or behavior; a statistically defined quantitative feature of some dimension of measurement; a species typical pattern of organism-environment relationships; and a social or cultural group–typical pattern of organism-environment relationships (Gollin 1984a).

Behind all three assumptions is the notion that knowing is ultimately an act through which a neutral mind discovers but does not mediate the structures of the world. Although it is a commonplace since Kant that such a position is indefensible, the old ideals and assumptions are so ingrained that their effects linger even in the work of some of those who are expressly committed to post-Kantian notions of inquiry as active. This is why the challenged assumptions often operate not only unintentionally and implicitly, but even at variance with an "official" commitment to their rejection. Because of this, I will look only at the ways in which "normality" actually functions in particular theories, and will ignore what the theorists explicitly say they are doing.

Organism and Environment Can Be Treated as Separate Entities

When theorists treat courses of biological and behavioral development as if they were largely independent of environmental variation and other sources of individual differences, they implicitly assume that normal courses of development are universal and immutable, except to the degree that unspecified factors influence the "rate" of progression or the level of final achievement along a predestined course. This concentration on observed regularities assumed to characterize all development makes it easy to neglect contextual differences (Gollin 1981; Toulmin 1981); as a result, observed regularities are mistaken for universals, which are then regarded as indicators of health and developmental adequacy (cf. chapter 2's argument that the definition of health involves evaluating how the organism functions in a developmental context).

One challenge to these assumptions is the "life-span" approach (Baltes 1979; Havighurst 1973) as formulated by Richard M. Lerner:

> From this perspective, developmental changes occur as a consequence of reciprocal (bidirectional) relations between the active organism and the active context. Just as the context changes the individual, the individual changes the context. As such, by acting to change sources of their own development, by being both a product and a producer of their contexts, individuals effect their own development. (1983a, p. 15)

This contrasts with ego-psychological formulations (Blanck and Blanck 1974; Hartmann 1958), which exemplify the universalist position. Here early development is presented as a reciprocal process involving a maturing individual and that individual's environment, especially as embodied by the mothering figure. Unfortunately, in the emphasis on the "autonomous ego" and the "conflict free sphere," the role of environment is undercut by placing "inborn ego apparatuses" inside the developing organism (Blanck and Blanck 1974). In addition, by positing a course of "normal" development in which the intrapsychic structures emerging in one stage act as the cause for health or pathology, ego-analytic developmental theory effectively removes the organism from its environment for the purposes of diagnostic and therapeutic decisions (Silvern 1984).

This move from a unitary organism-environment system to an independently functioning organism is accomplished by positing the "average expectable environment" (Hartmann 1958) and the "good enough mother" (Winnicott 1953). The latter provides the minimum requirements for the actualization of ego functions and their structures. As Blanck and Blanck put it in their classic and influential text, "Motility, intentionality, perception, and the like will follow the innately programmed developmental course, always provided that the mothering person is present and not so grossly interfering that she adversely affects the development of the ego and its functions" (1974, p. 31). The concepts of "average expectable environment" and "good enough mother" illustrate the way in which the first assumption operates in ego-analytic thinking, at least to the extent that these concepts hold the environment constant while attributing to the organism those properties that emerge from interactions with this "average" or "good enough" surround.

Even the Piagetian approach (Dasen 1972), with its emphasis on constructive transactions between child and surround, contains presumptions of a universal developmental course and outcome, along with the implicit notion that the normal course is the desirable course. Essentially, Piaget considers reason to be mature to the extent that it achieves a fixed set of Kantian concepts and forms that structure the world in terms of fixed objects located in space and time and subject to strict causality. The child does not begin from a neutral position in regard to the choice of alternative concepts, but is (genetically?) on course toward acquiring basically Euclidian notions of space, time, and matter (such as those notions of the matter "conserved" in all the famous experiments pouring water from one odd glass into another). The expectation that these structures are universal and cross-cultural rests on Piaget's assumption that they are rooted in the nature of the subject alone, rather than shaped by the dynamic between the subject and the surround. He also assumes that a selected set of cognitive categories—those central to the kind of analytic thinking he himself uses with brilliance—are the preferred mode of cognition for all modes of thinking; since he gives little or no credence to the possibility that other tasks and contexts might demand other modes of thought, intuitive and aesthetic modes of thinking come off rather badly.

All of these are examples of structural stage theories; in practice, if not in principle, they detach the course of development from both genetic and extragenetic influences. The only exceptions recognized in such universalist approaches involve genetic anomalies or environmental catastrophes. Save for these extreme events, development is treated as if it were independent of genetic or environmental variation (see Gollin 1981 for a more extended discussion of this point).

There are, of course, situations where it makes sense to treat "organisms" and/or "environments" as constants for the purposes of the research at hand. But in each instance it must be shown, not merely assumed, that doing so does not destroy the process of inquiry. Otherwise, the assumption that organism and environment are functionally reciprocating components will be lost. This is equally true in discussions of the modalities of moral experience, where sometimes the goals of the self (such as autonomy), sometimes the institutional demands of the culture (such as loyalty), are taken as fixed and given; the result is a failure to examine the ways in which changes in the self or the institutional niche are redefined by their relations to each other. This faulty procedure ignores the fact (of which the notion of exemplary judgment reminds us) that we are not dealing with fixed and static entities, but with continually revisable stabilities within ongoing and unfinished processes.

This inseparability of organism and environment (and, by extension, of person and the midworld) is nicely illustrated in the work of the ethologist Remy Chauvin (1977), who argued that only by systematically varying the environment within which the organism functions can the full range of organismic plasticity be elicited, and hence the full significance of behavioral variation be made intelligible. This entails a more accurate assessment of ranges of organismic capacity and depends on presenting organisms with novel as well as demanding environmental opportunities. Similarly, the challenges of radically different cultural niches reveal the unsuspected significance of "normal" human capacities and characters. Even science fiction—at least when it involves more than thinking up fancier ray guns—can be helpful in bringing out the implications of alternative worlds and challenges. For it often takes the shock in which the normal disintegrates for agents to understand the full potentialities and significance of their own taken-for-granted actions. The most terrifying examples of this are in the literature of survival in the Nazi death camps (Frankl 1963; Levi 1969); here the most taken-for-granted decencies become heroic achievements.

The employment of novel arrangements in this sense of the non-normative is fundamental to Baltes' insistence that we must include not only the normative age-graded influences that are central in traditional developmental psychology, but also normative history-graded influences that are "fairly general events or patterns experienced by a given cultural unit in connection with biosocial change," and that can "vary with historical time and can produce unique cohort-related constellations of influences" (1983, p. 95). An additional class of events to be considered consists of nonnormative influences on life-span development such as "environmental and biological determinants that, although significant in their effects on individual life histories, are not general" (p. 95). Such events as migration, divorce, death of a family member, and illness will contribute to the revelation of the plastic capacities of most individuals, and must be considered as relevant for all. With these sorts of considerations we move from taking the individual as a "case" of some general category, and begin to deal with an account, in dated time, which approaches the biographical—the realm of what Miller insists is the historical and distinctly human.

This insistence on individualizing procedures is also present in Kurt Goldstein's (1939) strategy of interpreting pathological symptoms as attempts by organisms to retain coherence, where coherence is always seen as the coherence of a specific organism with specific resources striving to function in an orderly fashion in a specific surround. This makes clear that pathology is unintelligible as a state apart from the processes—aimed at coherence—out of which it arises. The focus for understanding these processes is the organizing mode of coherence. This is equally true within clinical psychology, where one of the most fruitful ways of identifying pathological threats is in terms of their disruptions of the organizing mode of coherence found in the emerging narrative of the person's life. Similarly, threats to the moral self can be identified—in Kant-like terms—as those actions whose maxims would, if acted upon and generalized, contradict the coherent and responsible self whose maxims they are.

When we ignore the fact that what is "normal" or healthy for an individual involves what is "normal-in" a significance conferring context, inappropriate criteria and standards can slip in. The systematic methodological reasons for avoiding the uncritical transfer of the norms of one group to another—as disastrous for therapy and developmental assessment as for moral judgment—were spelled out elegantly by Silvern (1984), who emphasized the importance of including the fit between a child, or any organism, and the milieu in which it functions. To understand the nature and quality of the "fit" requires insight into the capacities of the child and the resources available from the surround. The nature of this relationship is exemplified by human infants' preferential attention to visual patterns that resemble faces, and their responsiveness to sound frequencies in the range of human speech. Here any understanding of attachment behavior (regarded as a reciprocal event between an infant and a caretaker) must be based on a detailed knowledge of infantile receptor-mediator-effector properties and those environmental features that fit these organismic properties (Gottlieb 1985). From the point of view of the other chief actor in this scenario, the mother (caretaker), the infant's qualities constitute the extra-organismic pattern of characters. The evaluation of the attachment relationship in terms of health, deviance, or any other dimension must be carried out in terms of the interdefined system components that the attachment pair constitutes.

The concept of "fit" is also necessary for evaluating a child's task success, either with respect to clinical criteria of adequacy or developmental criteria of progress. Task success is dependent upon the degree to which a child acts in conformity with the demands imposed by important life settings such as school, peer relationships, and home. Within these child-environment systems, "task difficulty" is an interdefined property of a child's resources (capacities) and of expectations (demands) of important persons in the child's world. As a result, the quality of task accomplishment is a systems property whose systems components include a child, other important persons, the demands on the child, and the child's resources and skills with respect to the demands (Silvern 1984).

In the same sense, health as a task accomplishment is a property of the organism-environment system, not of either in isolation. And if, as I have been arguing, moral action is an emergent in the same process, then morality will equally be a property of the person-midworld system, not of the person or institution in isolation. It is as futile to judge the virtues and vices of character apart from culture as it is to judge an organism's strengths and weaknesses apart from the demands and opportunities of its environment.

Similar Outcomes Are the Result of Similar Processes

This misleading assumption, like the first, is particularly evident in theoretical and treatment models which hold that some particular "normal" organismic state exemplifies optimal qualities for an assigned developmental stage. Just as the first assumption involves the neglect of contextual considerations, the second involves neglect of the processes by which designated outcomes appear; single-minded focus on the observed status of the individual and fixation on performance or outcome obscure the transactions that have mediated their emergence. Since both assumptions reflect the same nonevolutionary, nondevelopmental way of thinking, the same examples can illustrate either one.

Their interconnection is displayed in concepts of equifinality and multifinality (Silvern 1984; Wilden 1980), where the former term suggests that the same end can be attained by various means, and the latter that the same means can lead to different ends. This mix is illustrated in the syndrome of "hyperactivity," an ostensibly identical outcome that in fact may reflect a variety of etiological factors (equifinality); yet the presence of the same etiological factors does not necessarily lead to manifestation of the syndrome (multifinality). But even though the complex of symptoms included under the term "hyperactivity" may arise from multiple sources, they are often treated as if they made up a unitary phenomenon calling for the same clinical strategies regardless of causal agency. This is in spite of the fact that treatment plans suitable for one hypothesized route to symptomatic expression may be quite inappropriate to another (Gollin and Stahl 1985).

For example, central nervous system damage (so-called minimal brain dysfunction) is often taken as the cause of hyperactive behavior, then the children treated by prescribing stimulant drugs or a program of behavioral management. These substantive interventions are made in spite of the fact that the various behaviors thought to make up the syndrome of hyperactivity reflect a wide array of causal agents for which drug treatment or behavior modification procedures are not only irrelevant but obscurant: either can easily mask and miss causal conditions that lie in the family or in the classroom rather than in the child.

Some adjustment of family or educational setting for the child might be more ameliorative than dosing the "sick" nervous system or trying to get the behavior under instrumental control. Indeed, the drug-induced changes or the managed behavior modifications may well make the problem impervious to correction. In any case, the likelihood of effective intervention will be increased if we make differential distinctions between the ways in which the different factors condition each other in an open system transaction (see McNew 1984 for a discussion of the difficulty of attributing hyperactivity or learning disorders to neurological or any other factors; on research design see Baxley and Leblanc 1976).

The fact that analysis of the causes of hyperactivity is the beginning of its treatment makes the point that the relationship between process and outcome is prospective as well as retrospective: any behavior that is an outcome of a past process is itself the beginning of a further process. This temporal dimension is somewhat masked by the fact that we speak of the causal conditions of the past in verb-terms denoting process, but reify "the outcome" as fixed and noun-like. But which aspects of process are dubbed "outcome" is at least partly a reflection of the salience and values they have within our research or intervention; after all, behavior does not present itself in a neutral datum given as outcome, but is *taken* to be an outcome that has significance from some point of view.

This temporal dimension is so central that the connection between the first two assumptions can be stated in terms of it: the inseparability of subject and object reflects the simultaneous aspect of the process of their emergence; the sequential aspect in this same process is reflected in the fact that its outcome is inseparable from the process out of which it emerges (and from the further process which it begins). In spite of this closeness, though, there is a substantive difference in the ways in which the two assumptions are normally regarded. The idea that the identity of persons involves reference to context seems threatening to the sense of autonomy traditionally associated with agency; similarly, our commonsensical (and Newtonian) notion of objects as distinct lumps of matter seems to exclude the notion that their identity changes across contexts.

In contrast, the notion that the same outcome can be reached by dif-

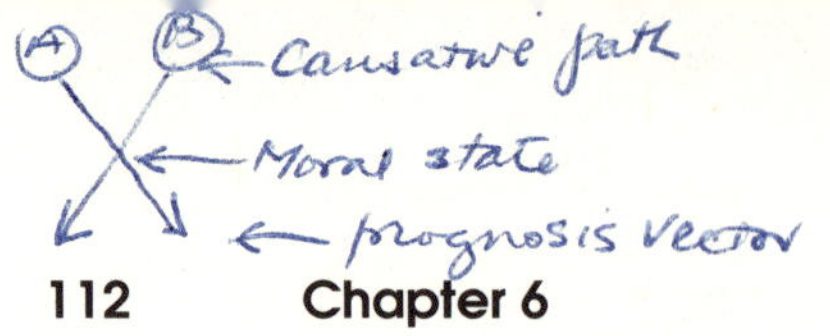

ferent routes is less disturbing. However much we are encouraged by the normality of the familiar to ignore it, the notion that outcome and process are connected is all around us: evolutionary outcomes like reproduction or the absorption of food can notoriously be reached by different—even bizarre and extravagant—routes. In terms of normative issues, even ordinary language recognizes Kant's point that the "same" act has a different significance if done from different motives. But the clearest awareness that the significance of outcome is a function of process is in our notion of medical diagnosis: a *sound* diagnosis is precisely one that differentially discriminates which of a number of possible processes has led to the symptoms as outcome. Indeed, it is only when symptoms can be grasped as the outcome of one disease process rather than another—placed, we might say, in etiological space and time—that they can be "run through and held together" as a specific disease entity; and the claim that they are *correctly* conceptualized as this disease depends on the success of the prognosis that follows from this analysis. It is perhaps laboring the obvious to add that this sequential story is literally embodied in the simultaneous interaction between subject and environment—and this at many levels, from that of the cell in its chemical environment to the organism in its surround.

Where the normality of the symptoms tempts the practitioner to rest in familiar significance and ignore differential diagnosis, the result is bad medicine and false prognosis. There is, as Professor Louise Silvern pointed out to me, even a danger of this in the "bible" of clinical psychology, the *Diagnostic and Statistical Manual,* where clinical syndromes are rather straightforwardly based on lists of manifold symptoms. The same confusions lurk in any analysis of moral "syndromes," such as states of character, in which no differential diagnosis is made of the different paths through which they are achieved, or of the different consequences they have in different contexts.

Science Is Value Free

That this assumption is a dead end goes almost without saying. Unfortunately, it can also go almost without thinking about the reasons why it is so, and what these reasons imply about both the nature of science and of value. Hence more is needed than a perfunctory acknowledgment that valuing is involved in selecting topics for research, or that the search for knowledge takes knowledge as a value. For once these facts are noted, it seems plausible that scientists could return to doing "hard science," leaving the (now duly recognized) values to chance, or to socialization, or even to God—in any case, letting someone else worry about them.

What we must get clear is how the value dimensions of science fit into the overall dialectical process through which self and other come to be, that is, how doing science is part of the humanity-defining process of articulating and revising purposes within history. The first step is to show how the assumption that science is value free is so closely related to the first two misleading assumptions that their denial entails its rejection. For if the notion of a value-free science depends on unacceptable assumptions, then science will not lose its authority because it involves values.

Historically, the assumption that science—or any other form of knowing—is value free rests on the claim that its fundamental statements can be seen as unmediated reports reflecting nothing of the observer's point of view. But this is incompatible with the rejection of the first two assumptions, which jointly entail that the world known or acted in is an emergent outcome, not a neutral given. Both that which is "taken" as a salient element within a description of experience, and the world that emerges out of extending such descriptions, must be understood in terms of the processes through which they emerge. These processes involve the simultaneous codetermination of subject and object, and the sequential determination of process and outcome, both of which involve a continual sequence of evaluations.

Consider, for example, what the simultaneous relationship between subject and object means for the process of apprehending something as a datum: a datum is something *given* only so far as it is *taken* by a knower as salient within the framework of organization that has arisen in response to her problematic situation; this framework reflects the value-laden interests whose frustration made the situation problematic, and whose satisfaction will constitute its resolution. Similarly, the person who takes the datum up into this framework *is* who she is as a knower, at least in part, because she is the knower *of* this datum. Her being a knower and the datum being a datum are not intelligible apart from their simultaneous relationship to one another.

The same links hold when we consider the datum and the knower in terms of the sequence out of which the knowing experience emerges: the significance of the datum is inseparable from the process of inquiry out of which it has emerged *as* a datum for a knower. As Dewey pointed out, it is central to the experimental method that "sensory qualities have their cognitive status and office, not in and of themselves in isolation, or as merely forced upon attention, but because they are the consequences of definite and intentionally performed operations" (1929, p. 113). Thus the sensation of blue has one significance if it is the outcome of a series of chemical operations on litmus paper, quite another if it happens to reflect the sky. Equally, the test of whether the significance of the datum is merely apparent depends on whether it can be incorporated into the ongoing process of inquiry and generalization. Once we reject the possibility that inquiry can be measured against a neutral standard not itself mediated by inquiry, then the judgment that certifies that inquiry is responsible must shift to the *processes* of revision *within* inquiry: "In every scientific undertaking there is passed a constant succession of estimates; such as 'It is worth treating these facts as data or evidence; it is advisable to try this experiment; to make this observation; to entertain such and such a hypothesis; to perform this calculation, etc.'" (Dewey 1929, pp. 261–62).

The criteria for being "worth" treating as evidence will themselves reflect concepts and methodological structures of the paradigms that shape the inquiry. As a result, the values informing paradigms will affect the range and type of data gathered, as well as the criteria in terms of which the "fit" of the theory to the data are determined. Hence the issue to be decided is not whether value-committing paradigms will be involved in all scientific investigation, but whether the choices made in constructing and testing paradigms are responsible and self-critical.

Part of this responsible self-criticism involves distinguishing between the concepts that embody the distinctive priorities and modes of organization of the chosen paradigm, and those "prejudices" that reflect the distortions of uncritical subjectivity. Some cases are easy (homophobes reviewing gay service records), but in others the criteria of relevance are themselves being reevaluated within the process of inquiry. Suppose, for example, one were developing a hypothesis about the relationship of genetic factors in sexual orientation to cognitive or physiological functioning: whether this move represents the intrusion of sexual stereotypes into research, or a potentially valuable insight into the relation between genetics and functioning, must wait on the fate of the research as it is tested in the critical community. This means that (to use the Kantian language of chapter 2) the judgments of relevance are not merely determinant, referring to a fixed and preexistent standard, but have an exemplary element that calls on the emerging critical community for concurrence.

The Concept of "Locus of Significance"

This chapter began by listing six different senses of "normality"; I have tried to show some of the confusions that result when they are used uncritically. Each of the senses is perfectly legitimate in some specific context and under some specific set of assumptions and limitations. Unfortunately, the term "normal" does not (normally) trigger the requisite questions about context and limitations. Quite the opposite. Hence using one static term to specify so many different context-bound relationships inevitably generates confusion. This can be decreased if, in regard to any claim that a behavior or a course of development is "normal," we specify the focus of attention in terms of which it is taken to have that significance. For nothing is normal *simpliciter,* but only in relationship to some point of focus. In this sense, "normal" functions like a relational term, such as "between," and is conceptually incomplete without a specification of the conceptual axis to which it is related.

The subject of all of this concern is not just the notion of normality, for this is only one instance of the ways in which static and acontextual categories can distort thinking and research. The real issue is the consequences the rejection of the three methodological assumptions have for attributing significance in any field, not just in the developmental sciences; these consequences can be summed up in the notion of "locus of significance."

An informal characterization of the notion begins with the fact that any given mode of transaction can be sustained only by a limited range of subjects and objects. As one goes up and down the evolutionary scale, the range of these subjects and objects varies enormously; what is meant by transaction is equally open to variation, and ranges from mechanical interaction between two elements to the sort of open system mutuality distinctive of moral relations between persons. Understanding any of these widely ranging elements is only possible if we fix in on that point of interaction where their varying significances are displayed, and apart from which they are unintelligible. This is the locus of significance.

Concentrating on the organism-environment range as an example, this can be put more systematically as follows: the methodologically primary locus of significance is in those meaningful, distinctive, and important perspective-derived macroproperties of a transaction whose components consist of the properties of an organism and of the resources and constraints afforded by a niche; these components constitute a range of elements dependent for their significance on some relevant macroproperties. The components consist primarily of those plastic functions of the organism that enable it to maintain itself as a subsystem in the face of perturbations, and those features of the niche that constitute either challenges or resources relative to the organism. Niche resources are interdefined with organismic properties; both function as components of a larger life-system, an ecosystem. Physiology and behavior are durable macroproperties of the configural patterns that characterize the living system.

Transition to Issues of Development

The rejection of the first misleading methodological assumption, that organism and environment are separately definable entities, means that their definitions must always reflect that process through which they codetermine one another. No range of organismic functions or range of resources or deficits can be considered normal or abnormal independently of one another. We cannot talk about the "good enough mother" apart from some changing range of children's needs, or about "task difficulty" apart from the child's capacities and the relevance of the task, or about "normal" body chemistry apart from the spectrum of demands that the body must meet in some resource-rich or resource-poor ecological niche. In each instance, organism and environment are constituents in a configuration that is salient within some observed locus that specifies the functional integrity of an operating system. Methodologically, the focus is on an activity such as "nurturing," for it is in relation to what sustains or distorts or destroys this activity that some dependent range of structures or activities are identified as significant.

The second methodological principle, that process and outcome must be understood together, is reflected in the attention that locus of significance pays to the consequences of shifting the focus of inquiry or intervention back and forth between the process and the outcome. Because an outcome is simply a segment of process on which we focus, we must be aware that, from another focus, each outcome is itself a process toward some further eventuality. Hence, except for limited and self-conscious research purposes, neither process nor outcome can be abstracted from the other, nor can either be taken as a self-contained entity.[2]

The third point, that science is not value free, is reflected in the recognition of the evaluative aspects of the judgment that some macro-property *worth* investigating is characterized by simultaneous interactions of complex variables, as well as ongoing and mutual feedback within a system functioning to maintain its own equilibrium (Thomas, Acebo, and Becker 1983, p. 653). Equally relevant evaluations may involve the role of system in survival, health, moral functioning, aesthetic satisfaction—indeed, any ongoing enterprise that structured the problematic situation within which the system became salient and worth investigating.

This notion of "system" is (as was chapter 1's discussion) based on the work of the microbiologist Paul Weiss, who arrived at his definition by analyzing a unit in biology as any composite retaining sufficient identity over time to warrant a name. As a conservative array of measurable properties, it contrasts with erratically changing "background" phenomena; as such, it is perceptible, detectable (measurable), and "significant" (1971, p. 9). This means that no unit qualifies as a system if it is a haphazard compilation of items or a rigid linkage of pieces or events. While units are predictable from information derived from their constituent parts, a system exhibits the opposite property: "The *state of the whole* must be known in order to understand the *co-ordination* of the collective behavior of the parts" (p. 13). A system has the unitary property of configural integrity over time and through space, despite limited perturbations of its components, although perturbations beyond a given range will produce a system of such different significance that it is a new system. In all cases, the relationship of the system's integrity in the transaction to the elements of the transaction will differ radically from those involving mechanical units. But these differences will only emerge if the methodological injunctions built into the locus of significance are heeded.

What we must now do is apply these methodological injunctions to increasingly complex and value-laden forms of transactions between subjects and others. For systems properties characterize experiencing at all levels, from particular aesthetic "takes" on a object as beautiful, to intentional acts, lives, and even history itself. If this is so, and if meaning and value consequently characterize these emergent macro-qualities rather than their elements in isolation, then understanding them will depend on sorting out their enabling conditions, that is, the relations between subject and other that are necessary to produce the emergent transaction that is the locus of significance. This will be of particular importance in analyzing (chapters 7 and 9) the ways in which the structure of human meaning (question 3) emerges as a systems property from the transactions defining the person's thrust (question 1) into her supporting or constraining world (question 2).

I have not argued in this chapter that the three methodological principles are new or even (with the possible exception of the first) particularly controversial, at least on the theoretical level. Rather, the point is that their implications have been insufficiently registered, even the developmental sciences that are their natural home, much less in such supposedly rarefied fields as ethical reflection. So if the reader's response to the principles is one of "of course," then so much the better, for this will make their extension to the areas of distinctly human behavior even more plausible. In this regard, Baltes' remarks on the metatheoretic status of life-span studies give an exact parallel:

> Indeed, none of the individual propositions taken separately is new, which is perhaps one reason why some commentators have argued that life-span work has little new to offer. Their significance consists instead in the whole complex of perspectives considered as a metatheoretical world view and applied with some degree of radicalism to the study of development. (1979, p. 7)

This sets our next task, which is to show that the metatheoretical world view summed up in the concept of locus of significance can be applied "with some degree of radicalism" to the study of distinctly human and historical development.

Chapter 7

Meaning as the Order of Processes

> The form indeed is "nature" rather than the matter; for a thing is more properly said to be what it is when it has attained to fulfillment than when it exists potentially.
>
> —Aristotle

The Continuity of Developmental Processes

We can now move from a discussion of "locus of significance" in terms of the general issues of developmental science to specific focus on the sort of behaviors specifically characteristic of humans—first to the processes of biopsychological development, ultimately to the kinds of actions that shape meaningful and coherent biographies and histories.

Tradition speaks of human life in terms of three ends, with morality the aim of action, truth the aim of knowing, and beauty the aim of appreciation. Rightly understood and revised, this division signifies more than the remnants of a faculty psychology: these are the ultimately inseparable aspects of the dynamic process of self-world articulation. The thrust of the self into the world (question 1) is shaped by the aim at morality; the order of the world presented to the self (question 2) is the aim of truth; and their synthesis (question 3) can achieve a beauty in which we can rest with appreciation. It is this process that gives the three questions their constitutional status, and this same process—with a wide range of elements—that unites all the levels of development.

Approaching the process through the notion of the locus of significance makes clear that at any point in it, subject and object are (simultaneously) intelligible only in relation to one another, that both of them are themselves the outcomes of process and the beginnings of process (sequentially), and that outcomes are identifiable as such only from within a point of view reflecting some problematic situation—itself an emergent from process and tested and refined in process. In order to apply these concepts with (in Baltes' words) "some degree of radicalism to the study of development" (1979, p. 7) so that it leads to parallels and generalizations about outcomes at the highest and most distinctive levels of human functioning, I will focus on incidents from the ancient war between nature and nurture. These classic disputes about whether the source of form is in the subject or the object are rich in implications, not only for their more technically dense incarnations today, but for disputes about the locus of form in other fields, including the moral.

Deprivation Experiments, Both Biological and Moral

A recent piece by Timothy D. Johnston, "The Persistence of Dichotomics in thc Study of Bchavioral Development," points out that in spite of the fact that

> the inadequacies of dichotomous views of behavior development that oppose learned and innate behavior, or genetic and environmental determinants of behavior, have long been recognized . . . they continue to exert a powerful influence on current thinking about development, often by way of metaphors that simply recast these old ideas in a more technical vocabulary. (1987, p. 149)

Johnston focuses on Konrad Lorenz's famous deprivation experiments, and argues that the dichotomies and distinctions made there continue to confuse us. Lorenz chose some species typical adaptive behavior, then asked how it could have been learned; he then deprived the animal of what he took to be the relevant opportunities for learning. If the animal still displayed the behavior even in the absence of any learning experience, he reasoned that the behavior was genetically controlled and innate; if the absence of the experience resulted in the elimination of the behavior, the behavior was learned. As Johnston notes, "Given the conceptual dichotomy between instinctive and learned behavior, the logic of the deprivation experiment is compelling: Deprive an animal of the opportunity to learn, and the only behavior that will develop is instinctive or innate" (p. 152).

In an influential criticism of Lorenz's work, Daniel Lehrman argued that "the assumption which underlies Lorenz's approach to the neurophysiological basis of behavior is that the neural events underlying behavior patterns must somehow be isomorphic with the behavior itself" (1953, p. 350). In other words, order in behavior is rooted in the (separately definable) order in the subject. But in the absence of any coherent story about how the genes—as the structuring element in the subject—*contain* behavior, the simple attribution of behavior to "the genes" does not constitute a meaningful explanation, for this "throws no light on its *development* except for the purely negative implication that *certain types* of learning are not directly involved" (p. 345).

Lorenz countered Lehrman's criticism by saying that the determining influence of the genes was transmitted to behavior as "information." The animal's adaptation to the environment is controlled either by genetically encoded "phylogenetic" information or by such "ontogenetic" information as the animal can pick up from the environment (Lorenz 1965). But the failure of the animal to display the target behavior after being deprived of sources of learning information does not always mean that the normal source of the behavior was learning, for the same interventions that eliminated the learning may have resulted in such "bad rearing" that, even though the innate programs for the behavior were present in the genome, they were blocked from actualization. This raises the nasty question of how to distinguish between these cases, for both involve a change in the environment that has produced an alternative state in the animal's development. Both are "information" in the relevant technical sense (Johnston 1987, p. 156); and in both there is the "same" outcome (the elimination of the target behavior), although each arises out of a different process and has a different significance.

A comparative analysis, such as is called for by taking the behavior in question as the locus of significance, might uncover some of the codetermining factors involved here, but this is not the course that Lorenz takes. Given his focus on species typical adaptive behavior and his commitment to an exclusive and exhaustive dichotomy between genetic and environmental sources of information, he simply assumes that some genetic form of the animal and some course of rearing are unproblematically normal. If so, then varying them independently will reveal what each uniquely contributes.

This is a tempting move for an evolutionary biologist who is properly and primarily interested in the ways in which natural selection has led to forms of life that have adapted to some environment. After all, such an observer's "perception of a developmental process is wholly dominated by his pre-knowledge of the outcome (i.e., of the adaptive form of the fully developed behavior or structure)" (Lehrman 1970, p. 45). But while it is both unavoidable and sensible to be interested in such real and actual outcomes, this should not mask the fact that the selection of any activity as salient and important reflects a value commitment on the part of the observer. As Lehrman points out,

> The investigator who wishes to understand developmental processes analytically must . . . have quite a different attitude toward the "normal." For him, the normal and the abnormal environment are simply two ways of treating the developing organism, and it is precisely by considering how the development is changed by any particular variations in the environment that he arrives at some understanding of the mechanisms of development. It would be overstating the case (but not by very much) to say that for the student of development, the normal environment and the normal path of development are no more meaningful or significant or perceptually prominent than any other environment or outcome. (1970, p. 339)

There are some nice parallels here to the ways in which moralists are tempted to think of agents and situations as so unproblematically normal that they can be discussed in terms having no relation to any particular time or context (else they would not speak so blithely of the problems of agent *X* being obliged to do act *A*). But as Miller would say, this hardly puts action into history. Nor does it lead to the kind of careful comparative analysis able to tease out the differential conditions under which—and only under which—the behaviors taken to be normal normally have the same results.

Lorenz has a further problem, with equally interesting implications: within his assumptions about the essential "givenness" of the normal, he can develop no criteria for differentiating the outcomes of "bad rearing" from those of successful deprivations, and must ultimately fall back on some intuitive notion of what the experienced ethologist takes to be normal and abnormal. As a result, an arbitrary element slips into any decision classifying a particular change as due to the interference of bad rearing rather than the lack of learning information.

A similar difficulty plagues disputes between those theorists who see changes in children's behavior as due to learning in response to tasks set by the environment, and those who see them as a sequence of stages rooted in the children's nature: where the stage-related behavior does not appear at the expected age and setting, stage-development theorists distinguish between a "competence" that, like the genome, marks true development, and "performance" that may or may not occur, depending on rearing and stimulation. But without some criterion for distinguishing one from the other, the claim that competence is present even in the absence of performance is simply a self-protective methodological device for excluding contrary instances (Fischer and Silvern 1985).

There is an equally instructive parallel in the perennial quarrels between those who attribute moral behavior to some source of moral competence in the nature of the agent (conscience, the moral will, something like a "moral genome"), and those who see such behavior as learned response to challenges set by the environment. Unfortunately, the dichotomous way of framing this question does not lend itself to resolution here any more than it does at the biological level. Thus when a political theorist (such as Hobbes) makes a "deprivation experiment" by thinking away all the learned behavior so as to isolate the "natural man," there is no way of deciding whether the aggressive animal he "finds" is the result of eliminating the conditions under which natural competence for goodness can perform, or whether it illustrates the fact that without learning there is no such behavior. If, on the contrary, his thought experiment were to yield a picture of an uncorrupted and noble primitive, he could as easily argue that civilization itself is such a case of "bad rearing" that our "normal" species behavior, full of wars and viciousness, is not morally adaptive. We would do better to abandon the whole enterprise as based on a misleading dichotomy (of this more later).

The Inseparability of Subject and Object

In all these disputes the failure to insist on the codefinition of subject and object results in the persistent tendency to think of entities—whether the genome or the moral agent—as having a nature and significance that can be stated independently of their specific functioning in some range of contexts. Only this false assumption makes it plausible to separate questions about the structure of the entity from those about how it functions in a given context; as a result, the behavior, although somehow "caused" by the structure of the organism, is thought of as externally related to it. With this view even Kant agrees: "The essence of things does not vary with their external relations" (1785, p. 439).

But consider what this means in the case of Lorenz's dichotomy between sources of information in "the genes" and in the environment. He is in principle committed to the notion that we can talk about the structure of genes quite apart from any context of functioning. But this is not coherent, for genes *always* have some context of functioning, usually the chemical context within which they produce the sequence of amino acids making up the basis of proteins; but their context is *never* those aspects of the world that are environment relative to the doing and undergoing of the fully constituted animal. Whatever genes do and however they do it, they never *directly* inform behavior. As Lehrman puts it:

> The problem of development is the problem of the development of new *structures* and activity *patterns* from the resolution of the interaction of existing ones, within the organism and its internal environment, and between the organism and its outer environment. At any stage of development, the new features emerge

> from the interactions with the *current* stage and between the *current* stage and the environment. The interaction out of which the organism develops is *not* one, as so often said, between heredity and environment. It is between *organism* and environment! And the organism is different at each different stage of its development. (1953, p. 345)

Put the matter this way: deprivation experiments may *change* environmental conditions but cannot eliminate them, for at every stage of development and for whatever unit (gene, organ, organism—or moral agent) one chooses, functioning occurs always and only in some relevant context. Thus "the important question is not 'Is the animal isolated?' but '*From what* is the animal isolated?'" (p. 343). Since it can never be isolated from *all* experience and context, the notion of the functional significance of the animal in isolation is empty (cf. the systematic unknowability of Kant's object "in itself"). If one takes the phrase "animal itself, its nature" to mean the animal apart from all context, then there is no such thing.

Lehrman's replacement of the gene/environment interaction by an organism/environment interaction not only makes his point that levels of growth and environment are integrally related, but reflects his focus on a particular level of analysis: given another set of problems and another discipline, a gene/environment interaction could be equally central—although only as long as the "environment" in question was that which was defined to be such relative to the functioning of genes, rather than appropriate to the phenotypical animal. Similarly, if we focus on a level of inquiry rooted in the problematic situation defined by our three original questions, the interactions of significance will be between persons and appropriate others. Here, too, the levels of entity and context must match: as I argued in chapter 1, no sense can be made of persons in a world defined solely on the level appropriate to organisms. The trick is to work out the nature of the transactions within which the various levels of stable entities are articulated, for it is only in the common structures of systemic modes of activity involving disparate elements that any coherent order will be found.

Where these salient transactions involve the production of amino acids, the relevant powers will be gene sequences as structural stabilities, together with whatever chemical environment is necessary for them to operate; where the indexical macroquality is, say, some pattern of birdsong (Johnston 1988), then the relevant elements will be some bird(s) in some niche. And where the focused-upon activity is some form of the distinctively human and moral, what is marked out are agents vis-à-vis other centers of moral dignity. This is the point in my process interpretation of Kant's matching of form in the subject and form in the object: the emergence of the human (Miller: the historical) entails the simultaneous emergence of a world of moral others and institutions within which it is unconditionally imperative to "treat others, and thine own self as well, always at the same time as an end, and not as a means only."

To sort out the relative stabilities making up the range supporting whichever activity is the locus of significance, whether birdsong or acts of duty (or some paradigmatic way of thinking, such as "scientifically"), will involve the kind of comparative and contextual analysis embodied in Lehrman's notion of a "selective rearing experiment." This entails a controlled manipulation of the variables governing the outcome of behavior in some niche. What is excluded is holding constant "the" (assumed to be normal) subject or "the" environment selected without reference to a determinate range of subjects. The aim of the experiment, in Johnston's words, is "to test a specific hypothesis about the contribution of a particular aspect of the environment to the development of a particular behavioral characteristic, not to distinguish between two different categories of behavior" (1987, p. 153).

That this is called a selective *rearing* experiment does not mean that all changes are brought about by "rearing" as opposed to genetic structure, since the point is to reject this very dichotomy (Lehrman 1970). It does mean that the impact of any intervention must be assessed in terms of its impact on the total outcome of the process within which animals with a particular heredity are reared within a particular environment. As Lehrman says, "Nature selects for *outcomes*" (p. 36), but not by causally affecting heredity *as opposed to* environment, or the reverse. Change takes place in the "continuity and interpenetration between the processes of growth and those of the influence of environment" (p. 36). Here the sequential process of growth and the simultaneous influences of environment intersect at an outcome taken to be the locus of significance for investigation. Hence any constancies attributed to these activities-in-environments will be properties only of these interpenetrations, not of organism or environment apart from some mode of transaction.

Even where some event, such as the impact of radiation on the DNA, seems to be limited to an element that can be isolated within the process of rearing, we are dealing with the behavior of a subject-context system, for what the genes *are* as powers to sustain some outcome will vary with their codetermining context: the same change in the genome (or in the character of the self) will have quite a different significance given a different context (or a different set of ethical challenges). Thus, commenting on a series of experiments, Lehrman concludes that "the behavior patterns concerned are not unitary, autonomously developing things, but rather that they emerge ontogenetically in complex ways from the previously developed organization of the organism in a given setting" (1953, p. 345).

As an illustration of these interrelationships, consider R. M. Cooper and J. P. Zubek's (1958) experiments comparing the number of trials needed to perform a learning task for two different populations of rats, one selectively bred over a number of generations as fast learners, the other as slow. Under isolated and impoverished conditions, both need essentially the same (large) number of trials to reach performance; in the intermediate condition of a "socialized" environment, the learning curves diverge widely, with the fast learners performing maximally after only a few trials, while the slow group still takes about as long as in the impoverished environment. But in an enriched environment the differences between the two strains again become essentially irrelevant, and both groups work up to task with minimal trials. Clearly, the significance of the fast learner/slow learner genetic strains cannot be stated independently of the conditions of thc trials.

Lest one think that the conditions of learning can be isolated as the significant variable, consider (Cooper and Zubek 1958) the effects of a variety of conditions on learning in a population of genetically *similar* animals: if trials are given all together, then learning under the impoverished and social conditions is almost identical; only the enriched condition seems to make any difference in speeding it up. But if trials are spread out and distributed over time, only the impoverished conditions make a radical difference; performance under the social and enriched conditions improves at about the same rate. In each case, the salient learning activity can only be understood as a function of the animal-environment system.

There are imperative warnings here for analysis at other levels, such as thinking of some structured state of character as always and everywhere producing activity having the same moral significance. What, for example, would be the significance of a character shaped by unbridled lust in a culture that reproduced asexually? Nor can we assume that some stable structure of family or state or education will be equally productive of some ("normally") desirable sort of activity for all kinds of selves with any sort of histories. These dead ends are avoided if we seek the systemic uniformities of experience in the "interpenetrations" of subject and situation, not in their false isolation.

This idea is driven home at the moral level by taking the act as *exemplary* of a dynamic relationship between self and other that can sustain and further dignity. It is this relationship that the concurring community is called upon to institutionalize, not the particular forms of self and other that entered into the particular act. Consequently, all such talk about the shared order of moral experience must be—as Kant insisted—formal, that is, independent of the particular nature in isolation of either act or situation.

That the focus of significance is on the dynamic structure of transactions also follows from the fact that the entities in question, whether a gene string, a singing bird, or a keeper of promises, bring their histories with them. Each, as outcome, incorporates the stages of its emergence in its relevant environments. In Aristotelian terms, this means that the matter "out of which" the outcome emerges is never "prime matter" naked of form, but always some formed matter; and what counts as matter is always functionally defined as matter *for* some further stage. Thus in *De Anima* he insists that the significance of nutrition at the level of plants, where it is the distinctive form, is different than at the level of animal life; in animals the powers of nutrition function as matter qualified by the further forms of appetite and movement. So for an animal that moves about and must identify its food, nourishing takes on different meanings than for a stationary entity like a plant. Similarly, nourishment takes on a further significance when it characterizes the activity of intelligent and fore-thoughted animals. Thus to say that some animals are rational qualifies the meaning of animality, but not by mere addition of a further characteristic.[1] It would be tempting, but misleading, to describe this as a genetic and instinctive program for nutrition that is common to all levels of living things, but open and flexible enough to allow modifications by "instinct" and experience in the nonrational animals, or by thinking in the case of humans. But to take both forms of nourishment behavior as a manifestation of the same identical basic nature would muddle together the differential significance of the activities.

The Drive toward Reductionism

Lorenz provides a particularly vivid example of this tendency to conflate activities having a differing significance:

> The presence of an instinctive act also seems to be detrimental to the development of an intelligent process having the same function. At least, it is true of humans. To be convinced of the correctness of this statement, one has only to consider the behavior of highly intelligent men who have otherwise good critical faculties, when they "fall in love" to carry out the undoubtedly instinctive reaction of mate-selection. . . . Higher psychological development may occur without any reduction of the instinctive, innate members of a behavior chain. (1937b; trans. in Lehrman 1953, p. 353)

Lehrman discusses a number of examples in which Lorenz and his followers ignore the fact that "functionally similar behavior patterns may be effectuated through very dissimilar causal mechanisms" (1953, p. 351) and, as a result, fall into a "merging of very different levels on the basis of superficial similarities" (p. 352). One example is the way in which Lorenz typically denies that purposive behavior is the emergence of a level (stage) beyond that of joining together complicated instinctual sequences:

> In a man working with the motive of getting food, the behavior directed toward this goal includes many of the higher psychic performances of which he is capable; the "motive" (goal)—the instinctive act of "biting and chewing"—has become drawn back to the end of a long series of acts, without, however, thereby in any way denying its fundamentally instinctive nature. (1937a; trans. in Lehrman 1953, p. 355)

This way of thinking treats the human search for food (cf. Aristotle on the nutritive soul) as involving the same mechanisms and same significance as a frog flipping out his tongue under certain stimulus conditions. But, as Lehrman points out, "the actual complexity and variety, and situational relevance, of the sources of human motivation make such statements meaningless, not merely because human motivation is more complicated than that of the frog, but because it is qualitatively different in organization and development" (1953, p. 355).

It would be easy to pile up examples, but not to find one more extraordinary than Lorenz's lumping together of the behavior of the amoeba and the infant because both "move toward weak stimulation and away from strong stimulation (the amoeba as whole, the child locally)" (Lehrman 1953, p. 351). Lehrman points out that "the *mechanisms* underlying the response in the two animals manifestly must be very different," and that to ignore this leads to a "'comparative' psychology which consists of comparing levels in terms of *resemblances* between them, without that careful consideration of *differences* in organization which is essential to an understanding of evolutionary change, and of the historical emergence of new capacities" (p. 351).

These failures to carry out differential diagnosis (see "Similar Outcomes Are the Result of Similar Processes" in chapter 6) are matched by philosophers who ignore the differential "mechanisms" of motivation and lump together acts of prudence, unthinking ambition, and acts structured by a respect for the dignity of others. Here, too, similarities in outcome have led to the assumption of identical process and shared meaning; here, too, what is missed is the "historical emergence of new capacities," for what is at stake is no matter of mere complexity, but of structure and history—hence significance.

But this activity of the organism/environment system as it emerges in time is precisely what is salient in any truly comparative analysis. Subject and environment can be marked out as stable ranges only by comparing them as powers that can or cannot sustain some activity taken as the locus of significance of which they are the range. Here comparison is not something that one might, or might not, go on to do after the judgment has been made about the nature of subject and object; it is the judgment itself.

The same point can be made in terms of the principle that processes and outcomes (and the further processes that emerge from outcomes) must be examined for both equifinality and multifinality: since any locus of significance is always an outcome, one must always compare the processes (the ranges of subjects and environments and their orientation) that have, or do not have, the same significance. Furthermore, the very selection of some activity as salient, and thus marking out some range of subject and object, can only be made from a point of view involving a value commitment itself understood only in contrast to alternative commitments.

It may seem unnecessary to approach these points from so many directions, but any argument that (like mine) makes so much of the methodological congruity between the developmental sciences and humanistic inquiry is liable to be seen as slipping promiscuously from talk about rats to talk about rational agents. But comparative methodology immunizes against this by insisting on the ways superficially similar outcomes (acting territorially, selecting a mate) arising out of different processes can have a correspondingly different significance. Equally, it forces us to note the reverse, that is, those cases in which some systemic mode of order (a coherent and vital biography, a modality of moral life within which persons are centers of dignity) can be an order shared over a wide range of agents and environments.

Among the outcomes of process are all the differing modes of being human. These differences have significance not just for theory, but for action: people fight and die over them, and each of us must choose who we are to be—else we are merely accidental in who we are. But such choices can be made intelligently only if we understand the relationships between agency and other that sustain, or destroy, the different kinds of narratives. Only then can we act so as to put together a vital and self-revising life (which has what Albert Hofstadter calls "consummatory ownness"; see chapter 8). Such lives, Miller says, emerge when agents create a midworld within which they act to revise purposes within history. In such worlds, differences between ways of living are not just an unfortunate fact to be tolerated, but the condition of living as an unfinished being in an unfinished world. But appreciating this openness of process is possible only if we understand the differing ways in which process and outcome are matched. Only then can we see, let alone respect, the legitimacy of other choices in relation to the vulnerability of our own; only then can we see that others must make other choices if they are to remain in history as vital agents. Ultimately, the fact that we articulate different self-worlds through our exemplary acts is not a threat to the integrity of our different choices, but a sustaining *condition* of their authority and responsibility. Others must choose otherwise as a condition of their being other—and hence of us being ourselves.

What will *not* be found is what many have sought, that is, a substantive continuity among selves (some form of ethical substance) or some moral structure in the world in itself. On the contrary, what unites us as persons is the dynamic form of the processes within which we come to be, not the matter of our concrete choices through which these differences are created. It is this same structure of dynamic form that gives us the critcria for rcjccting somc modcs of putting together a life: in (somewhat) Kantian terms, such lives reject the principle of respect for the dignity of others and oneself without which a center of human coherence and autonomy cannot come into being.

Irreducible Outcomes in Development

An inevitable objection is that the moments of coherence and autonomy on which the whole analysis turns are illusory: there simply are no distinctive modes of behavior (moral and coherent lives) that cannot be reduced to a "case of" something more fundamental. Typically, controversies over whether something needs to be explained or explained away are not about whether the behavior *occurs,* but about its significance. Thus the egoist and the Kantian agree that there are cases of promise-keeping, but disagree about whether they must be seen as instances of acting out of respect for the moral law or as illustrations of some form of self-interest. Similarly, the dispute in child development theory is not whether a child can "decenter" by accurately identifying what some visual array would look like from another point of view, or whether she can sort triangles with triangles and red things with red. The fight is over whether to describe these activities as cases of incremental learning or as behaviors illustrating an aspect of some global "stage" in the child's development. Both this sort of stage, and that involved in the emergence of the ethical as an achievement, involve the emergence of rules for the organization of experience which incorporate diverse elements under one rubric. We deny that such global reorganizations occur by showing how all the relevant phenomena can be subsumed under categories of the "lower" level, usually learning (in the case of children's cognitive activities) or some version of self-interest (in the case of ethical skepticism).

In developmental theory, the clash is between two models, each of which gives a different significance to the same events (Reese and Overton 1970). The organismic-structural model focuses on "the structure of thought that lies behind the diversity of manifest behavior" and sees this structure as grounded in the unfolding nature of the organism. The mechanistic-functional model "find[s] development (or more commonly, learning) in manifest behavior which varies widely across diverse environments and functions" (Fischer and Silvern 1985, p. 618). In other words, such order as is found in behavior is a function of the essentially mechanical impact of the environment on a basically passive organism.

Choosing between these approaches in terms of evidence is problematic, since the "observations made by the methods of the one approach are easily discounted as irrelevant by the proponents of the other" (Fischer and Silvern 1985, p. 618). This occurs because those who see context and those who see structure use different research designs that "simply hold constant the context or structure, respectively, or they ignore it" (p. 618). Thus Piaget (1952 [1936]), the most influential figure in the organismic-structural tradition, distinguishes between development, which deals with the stages of organization involved in acquisition of real knowledge, and (mere) learning, an acquisition of contextual input that he compares to the circus tricks of animals. In the latter case, any observed variation can be assigned to learning, thus leaving the thesis of developmental stages untouched.

> According to this approach, then, the possibility of diverse sequences is ruled out a priori. The observations that most cogently argue for individual differences in kind of development sequence (rather than merely in rate of progress of stage of fixation) are considered irrelevant. The possibility of diverse end points is ruled out by the a priori stipulation of a single end point, which is described sufficiently abstractly to subsume diversity. Diverse sequences are ruled out by the structured-whole formulation and the proposition that the sequence itself is logically necessary. These assumptions . . . effectively preclude the detection of individual differences or environmental influences. (Fischer and Silvern 1985, p. 620)

This systematic methodological insensitivity to possibly disconfirming instances is shared, as one would expect, by the mechanistic-functional theorists who set up experiments aimed at detecting differences in behavior matched to differences in environmental stimulation. To this end,

> tasks or testing conditions are manipulated to demonstrate the variability of behavior in different contexts, or problem-solving strategies are assessed to demonstrate the variability of behavior across individuals. However, no analyses are done to detect consistencies in behavior across tasks for a particular age group or consistent differences between age groups. (Fischer and Silvern 1985, p. 623)

In general, both models proceed by taking a different pole of the transaction as unproblematically fixed and "normal." Thus the mechanistic model tends to treat all subjects (regarding competence, if not performance) as essentially equal. In the most methodologically egregious cases, such as B. F. Skinner (1953), the organism is treated as an almost pure potentiality to be impressed with form (cf. Locke on the "tabula rasa" and Aristotle on the impossibility of Pure Matter). In contrast, a "pure" destinational view in the organismic tradition treats differences in environment as irrelevant to the stage development of the organism, except for providing releasing conditions and pathological blockages. Neither sees comparative analysis of the conditions under which different combinations of subject and context produce different outcomes as necessary, since neither subject or environment is defined by these changes.

Unfortunately, however, only an analysis that systematically compares the ways in which child and context vary with each other to sustain different modes of activity can account for both consistency and variation in behaviors. Ideally,

> instead of identifying a closed developmental sequence of structures, the investigator seeks to specify how structures show stabilities and changes in different contexts for people of different ages. Instead of identifying the regularity of environmental impacts, the investigator seeks to specify how environments have similar and differing effects as a function of human structures and motivations. No cognitive assessment can ever be free of the effect of age-related changes in cognitive organization. Thus the findings of both the organismic and the mechanistic viewpoints remain relevant, and the prescription for new research is to combine the methods of the two viewpoints. (Fischer and Silvern 1985, p. 625)

This "combination," needless to say, is not a little bit of this and a little of that, but a systematic varying of the child-context system in terms of its power to sustain salient and selected modes of activity. Indeed, it is a selective rearing experiment. What emerges is that some kinds of activities are indeed stage-like and global for closely matched combinations of child and context, others are incremental and have little continuity across populations and situations. Sometimes there is an emergent stage, sometimes not; the matter cannot be settled a priori by theoretical methodological mandate. But the necessary empirical investigations can be made *only* if the dichotomous approach, both in concept and in the construction of experiments, is abandoned.

Ethological parallels to these problems are found in the question of whether specific behaviors result from the unfolding of the genetic nature of the animal (as the stages of development are thought to result from the nature of the child as organism), or whether they can be accounted for in terms of learned behavior (phylogenetic information) from the environment. One extreme would be the claim that "nothing but" genetic sequences need to be acknowledged, since these remain constant while environmental shaping and releasing vary. The opposing view would see all ordered behavior as the result of constancies of environmental shaping. Johnston summarizes the limitations of these extreme positions in a way that parallels the problems of the radically exclusive positions in child development:

> Lewontin (1974) pointed out with admirable clarity that estimates of heritability (roughly, the genetic contribution to phenotypic variation) depend critically on the *range* of environments and genotypes over which variation in the phenotype is assessed. If we hold the environment constant and vary only the genotype, heritability estimates will always be high (indeed, they will always equal one) because genetic variation is the only contributor, by definition, to variation in the phenotype. Similarly, holding genotype constant (by using pure strains, for example) and varying only the environment will always produce heritability estimates of zero because there is no genetic variation in the sample. Differences among phenotypes can thus vary between "completely innate" (heritability = 1) and "completely acquired" (heritability = 0), for the same character in the same population, depending on the *ranges* of environments and genotypes studied. (1988, p. 623, my emphasis)[2]

Taking the Outcome as the Locus of Significance

It is in these ways dichotomous thinking about the sources of behavior generates a methodological protectionism that keeps out evidence that might disturb the assumed dichotomy. Unfortunately, to the extent that significance of behavior is a function of the *ranges* on which it depends, no analysis excluding this interpenetration of the ranges will elucidate the relevant significances. These can only be seen if we replace deprivation experiments with selective rearing experiments that do not assume that outcomes are due exclusively and exhaustively *either* to genes *or* to learning. It is only if we compare the differential impacts that changes have on the history of the animal-in-a-niche that the shifting and controlling relationships between process and the significance of outcomes can emerge. Only this sort of differential diagnosis will mark the differing significances of outcomes as they emerge from different processes, and begin other processes. Failure to do this fixes on some one meaning of the behavior as univocally and statically normal, and leaves the description of its causal genesis subject to a certain degree of arbitrariness.

We saw this, however briefly, in the way in which Lorenz's exclusive and exhaustive dichotomy between sources of information left him no criteria for distinguishing between bad rearing and deprivation of relevant experience as the cause of altered experience in a deprivation experiment. But a more provocative illustration is given by K. W. Fischer and L. E. Silvern's remark that in a situation in which "organismic and environmental factors interact to produce behavior . . . virtually any factor that is classified as organismic from one perspective can be classified as environmental from another, and vice versa." Their example is that of an arousal state, something that "is typically classified as an organismic factor. . . . [However,] arousal is not understandable aside from an arousing context, and the arousing aspects of a context cannot be identified independently of knowledge about the individual. When children encounter a situation that affects them emotionally, their arousal level changes. The arousal level is actually a characteristic of the child-in-a-context" (1985, p. 639).[3] Where there is a dispute about whether the order and distinctive form of experience is due to the subject or the object, this point is pivotal. If the experience can be described either way, then the fact that it happens cannot decide the question of its origin. The result of ignoring this is an unending dispute that blocks the fruitful inquiry that a better formulated question could shape into empirically relevant questions.

For example, the traditional way of dealing with "moral emotions" such as love, hatred, anger, and envy is to simply assign their form to the subject and have done with it, taking the environment as providing merely the precipitating cause. But, as Aristotle long ago noted, this simply ignores the fact that the emotions are contextually directed to quite specific features of the world: "The emotions are all those feelings that so change men as to affect their judgments, and that are also attended by pain or pleasure. . . . We must discover (1) what the state of mind of [for example] angry people is, (2) who the people are with whom they usually get angry, and (3) on what grounds they get angry with them" (*Rhetoric* 1378a20). Compare, for example, what anger means when it is directed against a moral agent because she failed to fulfill a moral duty with what it means when it is a response to getting a cheap birthday present. Or compare anger against a person—for almost any reason—with being angry with your dog. Or note how anger changes if we learn that the person is mentally ill, rather than morally vicious. And if the subject is love, the whole of literature is testimony to the way in which its objects make it what it is (Solomon 1976; Sarbin 1989).

Not only will none of these complexities be intelligible if the form of the experience is assigned exclusively to nature or nurture, but all the woes that beset the partisans of that infamous dichotomy will plague any moral analysis that uses its methodology. Thus if morality as an irreducible achievement is explained by attributing its form to the subject's inborn moral competence (the conscience as moral genome), then it can be self-protectively defended against counterexamples by claiming that performance was blocked by bad rearing, lack of teaching, failure to be exposed to the releasing conditions of experience, and so on. The reverse view sees morality as the subject's unmediated intuition of a moral Form found in the world itself. Failures of performance can then be blamed on interference with the requisite capacity for moral intuition (moral blindness, distortion caused by bad rearing), not on the absence of essential moral Form in the object.

I do not recommend that we take this bloodless dance of alternatives seriously, but the opposite: nothing is less fruitful than the question of whether the source of order is the subject *as opposed to* the environment. To the extent that any particular experience can be attributed equally either to the self or the world, the nature of that experience cannot be used to help us decide between the alternatives. This is because the alternatives are logical extremes, rather than hypotheses about what has an impact on the "selected rearing" of persons in cultures. But if the order of moral experience is in the constancy of relationships (Kant: of mutual respect for dignity) that characterize the interpenetrations of different subjects and environments, then it cannot be captured by any analysis that holds either constant. Nor will any such analysis catch its *absence,* that is, those outcomes in experience that appear to be emergent stages but are quite explicable (as the skeptics have said all along) as self-interest behind smoke and mirrors.

An analysis that *will* not only catch the dynamic structure of these interpenetrations, but give us a conceptual vocabulary to make sense of them, is that of Albert Hofstadter, and I turn now to his work.

The internal transactions are highly variable:
mood/health/body state/energy/intellect/training
all contesting perpetually within the transacting persona.
The internal environment is in constant flux, stimulated by
itself and the external world concurrently in varying
proportions. Cf DAMASIO?

Chapter 8

Albert Hofstadter and the Dialectic of Process

> Form was the first and last word of philosophy because it had been that of art; form is change arrested in a prerogative object.
>
> —John Dewey

The Next Steps of the Argument

The themes relevant to our present concerns are found in Hofstadter's extraordinary (though neglected) contributions to aesthetic theory in *Truth and Art* and *Agony and Epitaph*. His analysis of the nature of art and its relation to other modes of consciousness expands the argument that neither subject nor object is intelligible apart from thc other, hence the focus must be on the qualities of their transaction. That analysis from such a different field bolsters our general theme is reassuring, but the fact is that I also need Hofstadter's integrative categories to pull together the increasingly diverse strands of the argument (locus of significance, exemplary judgment, the midworld and so on). With these in hand, we can come to some sort of closure with our three questions.

In its present formulation, question 1 represents the thrust of the self into the world; this stands as doing over against the undergoing represented by the thrust of the world onto the self, represented in question 2; only by taking their synthesis (question 3) as the locus of significance can the conditions of their mutual functioning be understood. It is on this last point that Hofstadter is particularly helpful, for his thinking illuminates the systemic qualities these human transactions must have if they are to be vital and choiceworthy modes of activity.

His analysis associates each element of this dialectic process with a form of truth: from the point of view of the person, the realm of doing is exemplified by truth of things, that of undergoing by truth of statement, and the synthesis that is the locus of significance by truth of spirit, or ownness. The final, integrative category is explicated in aesthetic terms, and art is taken as its most striking and available source.

This position contrasts dramatically with that of most twentieth-century philosophers, few of whom have thought aesthetics the "hard core" of the philosophical enterprise, or taken aesthetic appreciation to be anywhere near as fundamental as knowing and acting when it comes to defining humanity (this bias is reflected in question 1's focus on the *moral* self acting into a world, followed by question 2's query about whether we can *know* if this world supports the actions). In contrast, Hofstadter argues that aesthetic activity is a fundamental dimension of man's being, an integral way in which both self and world are articulated. For him, aesthetics is "near the very center of concern in philosophy" (1965, p. 1). This centrality rests on the fact that art gives us the most intense and complete experience of the kind of closure and harmony between self and other—where each is "own" to the other—that characterizes *all* experience when it is most distinctively and valuably human. For this reason, it can give us insight into the dynamics of process within all human transactions, including those moments, lives, and histories that are ultimately fulfilling and choiceworthy.

This position has suggestive parallels to Dewey's analysis of the aesthetic experience as the adverbial quality of any experience as cumulative and integrative (Dewey 1934a). Both thinkers take salient transactions marked by aesthetic quality as the locus of significance, and by doing so reveal the formal structure that must characterize the interpenetration of the self and other for this quality to occur.

The Historical Dialectic of Aesthetic Theory

Hofstadter begins *Truth and Art* with an examination of the ways in which the relation of self and world have been treated in the aesthetic tradition.[1] (The conceptual sequence he describes parallels the account of the development of *ethical* and *epistemic* thinking given above.) Once we recognize the "fundamental burden of the Kantian discovery of . . . the feature of spontaneity and originality of the human mind in all its enterprises," then it is clear that the truth of art (or, I would add, science or morals) "cannot consist in the mere copying of an antecedently given being" (1965, p. 13). Historically, the rejection of any such crude naturalism has led to theories that see art "as the expression of subjectivity," which is the projection into the world of that which is already clear in the self. But although this is *an* aesthetic that indeed "reflects how the essence of art took shape for a while in the actual life of man," it is not yet adequate to "the whole aesthetic-artistic life of man" in which "such phenomena as expressionistic creation and experience . . . are but one manifestation" (pp. 19, 21).

Recognition of this failure calls forth the "joint revelation theory," a synthesis whose aim is to "overcome the one-sidedness of an art and an aesthetic which are merely subjective or merely objective—or merely balanced between the two—by developing the idea of the *constant joint presence* of subjectivity and objectivity in art, allowing these factors to vary in their degree of revelation in the work according to their character, intensity, and degree of development" (p. 24). But this is still inadequate, for

> just as in expression theory inward reality already exists, and the word is the means by which it is brought to outward manifestation, so in the joint revelation theory the two-fold inner reality already exists and the word is merely an utterance of it. The so-called creativeness of the poet's intuition is, after all, only his capacity as a miner, to dig in both directions in order to uncover a lode already there and to dig out as much as he can; and his productiveness as an artist in regard to the work is only his capacity to carry up as much as he can to the surface of pigment or sound. In short, there is no genuine creativeness here at all. (p. 33)

We can take account of this creativeness by recognizing that art is a language whose truth is not a reflection of antecedent realities of *either* self or world (in whatever combination). Instead, it exemplifies the process by which all the levels of self and world come to be defined. From this perspective, within which "experiencing" serves as a general locus of significance, we can see that

> all language, whether in science, practical life or spiritual life, has the same generic character of meaning. Namely, all language articulates human being. Differences that separate languages . . . are due to the different ends toward which they are directed. Each such end is related to an ideal of truth. Hence it becomes essential to distinguish the different truth-goals connected with these different languages, so as eventually to reach a concept of the ideal that prevails in the domain of spirit, and of art in particular. (p. 52)

Since the end and type of truth differ for each language, so do the *criteria* of adequacy and evidence, and even the kind of entity of which truth is predicated. Man *is* the aim at these forms of truth, and this fact (metaphysical, but a fact nonetheless) give us criteria for evaluating the different languages through which he is articulated—including different lives and modes of living carrying out these different aims.

The Level of Truth of Statement

The first aim is that of theoretical truth, which applies most obviously to statements.[2] This is a quite standard sense of "truth," in contrast to Hofstadter's use of the term at other levels, where he speaks also of the truth of things and acts, and, ultimately, of the truth of man himself. But "if 'truth' seems to the reader too limited or too presumptuous, the term 'validity' might also be used. This is suitable particularly because it combines the connotations of rightness and cogency that belong to the adequacy of language as such, and consequently its use helps to keep our ideas about linguistic adequacy pointed in the proper direction" (1968, p. 90). But terminology is not the point. What counts is that all of the entities considered true (or valid) are so because they constitute an essential dimension of man and world process vis-à-vis each other—not because they reflect an antecedent reality. What is "true" is not simply an "ex-pression" of what was already definite though unexpressed (1965, p. 83): "On the contrary, the principle by which each symbolic form is form-giving is original, in the sense of originary; it is constructive of a meaningful ordering in its own spontaneous way" (p. 3). In J. W. Miller's terms, these modes of symbolism are functional objects that jointly construct the midworld, and the acts whose shape they articulate are exemplary in shaping the ongoing human transactions that are our histories.

It is because these aims represent the ongoing aspects of "originary" process that there are three aspects (thesis, antithesis, synthesis), which roughly correspond to the traditional division of "faculties" into knowing, acting, and appreciating. These three goals of truth can be described by variations of the traditional formula for truth as the "adequation of intellect to thing," and we can speak of theoretical truth, practical truth, and spiritual truth.

The theoretical level reflects the classical ideal in which truth of statement aims at the "adequation of the language to the thing it speaks about, in such a way that it says of what is that it is and of what is not that it is not" (1965, p. 92). Hofstadter gives this ideal the essential post-Kantian twist, insisting that what is "uncovered" is still partially a function of the activity of thought. Thus what is ideally absent—most especially and deliberately in scientific cognition—is not the active influence of all mediation, but the active structuring of thought that is *individual*. For the aim of this sort of truth is to be "objective, that is, the same for every self regardless of the self's own selfhood." This is an articulation of human being, but one that is "impersonal and selfless" and concerns "only the thin residue that is left when all individual selfhood has been removed" (1970, p. 104). Man is not man *without* this activity:

> Assertion is a form of man's being. It is a special form that human being takes on, namely, the intending of an entity in the way of aiming to uncover it in the way it is. Here there comes into play man's capacity to point to entities, to aim at an entity with the goal of striking it as a target of the understanding. This is not a mere capacity; man must exercise it, so that it is a necessity as well, or part of his "nature." In bringing this capacity into play, man brings about one form of his being as man. (1965, p. 99)

But he is not fully man with *only* this. The kind of "self" that is the necessary correlative of a world of impersonal entities located in a matrix of scientific space/time is the sort of impersonal self, not yet moral or aesthetic, which emerges from P. F. Strawson's (1959) minimalist transcendental analysis. From the point of view of full humanity, this is a "thin residue" indeed. It is far from the sort of being who could sustain, or even need, the coherent narrative of being-in-the-world that persons demand as the condition of their own identity. But it is an essential stage in the process through which self and world emerge, and represents an aspect of undergoing in which the self takes account of the emerging and codefining other. Science is only one aspect of this, albeit the most clarified and impressive sort.

The Level of Truth of Things

That theoretical self and truth of statement are only partial aspects of the human dialectic means that they have an antithesis in the "truth of things" found in the realm of practice, of doing. Here the self insists that the world adequate itself to the intentions of intellect and will. The two modes of truth can be contrasted in terms of the role they give to the self: at the level of statement, "the self . . . must become an intention identical with what is not itself. It must abandon any effort to be on its own account in order to be an intending of what-is in the form of what-is as such" (1965, p. 103). But in the truth of things, the rampant self "sets itself at the center of being" (p. 114), taking what-is as mere material to be shaped into a form true to the dictates of will. All manner of things, including persons, objects, and even theoretical statements, possess thing-truth when there is an "agreement of the thing's existence, or the thing as it is, with the concept of it as it ought to be" (p. 105). So when a statement "does what it ought to do, namely, signify that which is, it signifies as it should, and therefore signifies rightly, so that being right, it is true (now, in the second sense)" (p. 107).

The realm of thing-truth covers all that is traditionally considered by teleology and deontology; its limitation and dialectical incompleteness lie in the regress of judgments it unavoidably sets off. For the process in which the self shapes the world to its will and judges it by the self's own norms and ends inevitably contains an "open question" that asks of every end proposed or norm recognized whether it itself is good or justified. Ultimately we are forced to evaluate the person who wills: "To be man, he must attempt to govern things in the world. But in this attempt to set himself up as governor, he already subjects his own self to an ought, the ought-to-govern, that belongs to him as human. The will to govern *is* intrinsic to man as a radius of his being" (p. 121).

The Level of Truth of Spirit

An answer to the question of whether man is as he ought to be—whether he is himself "true" in the relevant sense—cannot be found in the practical sphere: "The question of what justifies the ultimate decision of the will is not answerable by another act of the will" (1965, p. 125). The regress can be broken only by a kind of knowledge that is the synthesis of the two previous forms. This is like the truth of statement in that it *presents* the self with a value; this value is, like the truth of things, the self's *own* truth and not an impersonal and alien fact. Although given to the self, it must be the self's own, yet not as the whim of the self, but as what is fitting and valid: "There must be in this truth (1) something of the nature of an uncovering or truth of understanding, like the truth of statement, and (2) something of the nature of a governance, i.e., something like thing-truth, practical truth, or truth of will. And (3) these cannot be unrelated to each other" (p. 130). The *dynamic* between these modes of relations is fundamental: a self that existed simply in the modality of doing could set no limits to itself, and could establish no measure to give limit to its intrinsically infinite will. Hence there must be an undergoing in relation to an other that sets some limit. But if this limit is not to destroy the authority of the self, it must not be simply external and indifferent to the self, but somehow the self's *own* limit. How can this be? How can this limiting measure be both like the act that needs measure and like the found limit that sets it?

One perennial answer is that the world and its measures of law were created by God precisely as fitting to the essential nature and destiny of the self; their common source assures their fit. This is not Hofstadter's move, nor is it compatible with the naturalistic assumptions of question 2 that "there are no ahistoric norms" that set limits to the thrust of the self. *Within* history, any norms will have to be grounded in the conditions for the emergence of vital and sustaining self-world transactions, not in the fact that self and world happen to fit—even if the fit were divinely arranged.

Consider Hofstadter's description of the conditions under which "governance by violence . . . [becomes] transformed into governance by consent." For this to happen, the cogency of that which persuades as a "rightful power" (1965, p. 132) must be grounded in a kind of dignity and personality akin to that at which it aims in me:

> Instead of imposing my will on the thing, I behave toward it in such a way that the thing also governs me. It does not govern me, however, by a violent imposition of its will upon me, as though I were a mere thing or instrument. . . . Rather, it extorts from me the yielding—the identifying of my ought with its ought. It is cogent with regard to me, it persuades me by a power it has to persuade one who has understanding and will; and on my part, as persuaded, I acknowledge it in its position of cogent governance. (p. 132)

This sort of truth is not alien to self, but an aspect of self in its otherness, for "I could not yield my will to the thing, identifying my concept with its concept (my ought with its ought), unless at the same time its concept were identical with my own" (p. 132).

This familiar theme from German idealism has already been discussed in chapter 2's analysis of the Kantian claim that man "is subject only to *laws which are made by himself* and yet are *universal*" (Kant 1785, p. 432). There I argued that this sort of universality was so abstracted from the motivations of any individual person that Kant's account of the motive of duty ("Reverence" for the law) was empty. But Hofstadter gives us an account in which the other is both a governance from "without" and yet is my *own* other with a command relevant to *my* history and motivation. Here the problem is not motivation, but the threat that the command becomes so much my own that it lacks authority over me.

As an illustration, consider the way in which "conscience" changed from its original sense of "knowing with" others, and became almost the opposite, that is, a judgment whose grounds are so particular to the individual that it sets her off *against* others. In this latter sense, the "voice" of conscience speaks in moving personal accents indeed, but the grounds of its authority become increasingly problematic. Such problems would remain even if one could somehow root conscience in the will of God, for even this must be shown to be nonarbitrary, and judged in regard to thing-truth. As Kant said, "Even the Holy One of the Gospels must first be compared with our ideal of moral perfection before we can recognize Him as such" (1785, p. 408; Hofstadter 1965, p. 113).

The root difficulty in all these ways of thinking is that self, as the source of form, is thought of as an entity, a substance, which has its being apart from the processes through which it comes to be. As a result, there is no end to the regress of judgments on the particular and contingent forms it happens to take. It is *only* if that which sets limits to the self is grounded in conditions for *any* vital and self-revising process of self and world that there can be an authority that gives measure to act, and a valid shape and style to will. The necessary structure of any satisfactory resolution, whether idealistic or naturalistic, is given in Kant's argument that it is only insofar as we recognize the dignity of *persons* that we will have an other that "imposes . . . a limit on all arbitrary treatment of them" (Kant 1785, p. 65). Hofstadter shows what this means within a point of view that takes the human task not as embodying a universal will (even if it is identical with our own), but as working out individual styles of self-in-the-world in relation to one another: "What we are trying to say is our own being, which has yet to be" (1965, p. 83).

The general drive of the argument is straightforward: the processes of self-world articulation will be sustaining and vital, characterized by aesthetic macroqualities of integrity and fittingness, only where self and world are in mutual harmony. Here what the world is reflects the self, and what the self is reflects the world. Self and world so interpenetrate each other, and are so much each other's own, that a quality of fittingness characterizes their interaction. As a result, the measure that the self meets and freely acknowledges is not a constraint that represses and denies it, but the condition of its fullness and completeness. This "*mutual* adequation of intellect and thing" (Hofstadter 1965, p. 130, my emphasis) is first achieved at the level of spiritual truth.

The Relations between the Three Modes of Truth

In *Truth and Art,* spiritual truth is developed implicitly as the form of truth called for by the incompleteness and partialities of previous aesthetic theories, and explicitly as the synthesis of thing-truth and statement-truth. But the *need* for this completion is ultimately rooted in the unstable processes of doing and undergoing necessitating the continual revision of ends as the price (Miller) of not being "banished from history." There is first (logically and temporally) the necessity that the self be separated off from the other, distinguished from and in terms of the other. But there is also the recognition that the other thus distinguished is, precisely *as* a condition of just *this* sort of self, the *own* other of the self. I will refer to the first level of process, that which generates the separation overcome by truth of spirit, as "individualizing ownness":

> This is the paradox of consciousness and the root of its living and ultimately spiritual dialectic. I am conscious and can exist as an ego and a person only insofar as I can reach beyond the barrier that separates me from what is other than me and have it as my own. In order to be as a conscious being, and therefore as the possibility of a knower, a free man, and eventually a loving and devoted spirit, I must exist in a condition of difference and otherness with what is nevertheless mine and to be mine. (Hofstadter 1970, p. 46)

While this level of separation of self and other is *necessary* for the "possibility of a knower, a free man, and . . . a loving and devoted spirit," it is not sufficient for its actuality. This is still the level of the "ownness of the other *qua* other" (p. 48), and the necessary separation of self and other (individualizing ownness) must itself be overcome, yet not abolished, if consciousness is to reach its "ultimate vocation":

> It is necessary to return to fundamentals if we are to understand the role of consciousness in life. Man's basic and universal need is to be with what is other than himself—and this includes himself as other, too—as with his own, or even more fundamentally, to transform other into own. Every mode and form of human life and activity is a particular way in which this basic and universal need is satisfied. All substantive philosophy has recognized this need as fundamental to the being of man, beginning with the Platonic Eros, Aristotelian entelechy, Christian love, Indian bhakti, Confucian harmony of heaven and earth, Taoist Way, and Buddhist Nirvana, and coming down to the concepts of freedom in German idealism and its offshoots, will and will-to-power in Schopenhauer and Nietzsche, appropriation in Marxism and existentialism, the I-Thou relation of Buber, the behavioral adjustment to environment, the Freudian libido, and the Jungian integration. Everywhere man tries to formulate for himself, whether in abstract concepts or in concrete images, the fundamental impulse of his being, it takes the shape of an overcoming of estrangement between himself and that which transcends himself, a transformation of his relationship to other into a relationship to own. (p. 49)

I think Hofstadter is right in this rather incredibly inclusive historical claim. But even if he is not, or if its plausibility can be defended only by making "ownness" so inclusive as to be essentially empty, this does not affect the essential point that "one fact is plain and unspeculative, namely, that man exists in the world, and that with the occurrence of the human mind there is given a clear possibility of a self-interpretation of what-is, in which what-is attains to meaningfulness and the possibility of truth and falsehood of its own being" (1965, p. 172). Surely there are, in plain fact, salient and seizing experiences that (as loci of significance) mark out those moments within process where self and other are in mutual harmony. I will call such passages of experience "consummatory ownness," to distinguish them from the level of "the ownness of the other *qua* other" (1970, p. 48), which is the process of individualizing ownness they presuppose. I apologize for these awkward and ugly names for stages (and there will be one more) in an essentially seamless process, but without them Hofstadter's argument is difficult to pin down.

In the earlier chapters, I worked from the difficulties inherent in some common questions toward approximations of both these notions of "ownness," particularly in two of the areas Hofstadter includes in his historical sweep, that is, "the concept of freedom in German idealism and its offshoots" and the "behavioral adjustment to environment." In the study of the latter, the developmental sciences have obviously focused more on untangling the sequences of development than they have on those moments in development that set the measure for value. Nevertheless, these sciences proceed at their peril if they fail to acknowledge the particular and distinctive "achievements" that occur within this process (this is the argument of chapter 1). Kant, on the other hand, is never in danger of reducing the irreducibly moral to the mere heteronomy of stimulus and response. In his work (or in a "Kant-like" position, if one prefers) the formal and nontemporal co-relation between self and other (the relationship between the transcendental object *x* and the transcendental unity of apperception, *or* between the dignity and freedom of self and that of others) has its analogue and history in the emergence *in time* of these very same modes of self and other. In all of these discussions, the concept of locus of significance is the inclusive rubric under which *both* levels can be understood.

Relationships between the Levels of Ownness

There is an apparent conflict between these aggressively naturalistic arguments and Hofstadter's teleological language—between, one might say, the discussions in developmental science and those in German idealism. Thus in *Truth and Art* he roots this "universal need" for consummatory ownness in a "model" of man as the "will-to-be" (1965, p. 149), and in *Agony and Epitaph* he speaks about the "vocation of consciousness" in the following way: "Consciousness exists *in order that* the ego should be able to differentiate its other from itself so that it can thereupon proceed to establish the intimacy of ownness between the two" (1970, p. 53, my emphasis). Lest the alarm bells set off (understandably) by this sort of language drown out the argument, I want to emphasize that Hofstadter's position is *not* a revived version of a voluntarist metaphysics. There is no point in the articulation of self and world where being suddenly attains "its own self-disclosure" and finds that ownness is not a vulnerable achievement but an inevitable given where "the truth of being that was here arrived at, though of the third kind, was also . . . a truth of the first kind, as though in beauty, goodness, and sanctity, being finally uncovered itself to itself in its pristine splendor" (1965, p. 172). At no point does he postulate a context of meaning independent of what emerges in history and act, for "to *be* at all is to be something so determined as a theme for meaningful interpretation" (p. 85). In Miller's terms, all meaning is part of the midworld. This is a point to which we will return, for it is of profound human significance that the only world within which we can be "at home" is that made actual through some participatory act. Yet it is undeniable, a "plain empirical fact . . . which metaphysical theories are challenged to explain" (p. 172), that within this actual midworld there *do* emerge consummatory moments that disclose the possibility of self and world being own to each other. Indeed, it is because no moments are more compelling than those in which art seizes the self that both *Truth and Art* and *Agony and Epitaph* present ownness primarily in its aesthetic mode.

Ultimately, I think Hofstadter's teleological talk reflects the fact that the creative being of man is never exhausted in the description of the way the past has been incorporated into the present. The time in which man exists *as man* includes the future, and his present cannot be described apart from that which is not yet. He is an "arrow pointed upward" (1965, p. 134) toward further stages of completion, further challenges and recompletions. In this context, the point of saying that "man's basic and universal need is to . . . transform other into own" is to deny all theories that see human action as aimed at a form fixed either by God or Darwin. It is to insist that man's essential incompleteness calls forth a *growth* involving stages of emergence and reorganization, not merely the accumulation of units. This is why even descriptive science must deal with man's irreducible "achievements," which cannot be evaluated in "determinant" judgments, but only in "exemplary" assessments projecting future states that requalify the meaning both of the present and of the past out of which it has emerged. The criteria for these assessments are grounded not in some fixed form apart from the process of growth, but in the *measures* within doing and undergoing that give the conditions without which self and other cannot come to be in mutual harmony. What is necessary and universal about these measures can only be captured in formal principles giving the structure of measuring and ordering, not in a prescription that tries to specify some separate and substantive shape of the self or the world.

The fundamental and pervasive nature of ownness is reflected in the fact that it is not a univocal concept, but is "subject to a universal ambiguity; the mode of union varies with the mode of being it unites." Thus it occurs not only as the "pervasive and ruling motif" (1970, p. 34) of human life, but exists in "analogous" ways even at the most primitive levels:

> All purposive behavior, even down to the tropisms of primitive plant life, consists basically in this, that the organism acts so as to impose its own aim on the materials to be dealt with. . . . Adjustment of the organism to environment and environment to organism, which is the central feature of life behavior and ecology, is determination of how the two are to belong to one another so that the environment should become the organism's world and the organism the world's inhabitant. (pp. 34–35)

As with any ultimate metaphysical concept, the danger is that indefinite extension will end in mere emptiness: applying to everything it says nothing about anything. What is going on, however, is that Hofstadter is taking those exemplary instances of ownness where "what it means, in the fullest sense, to belong and be belonged to, is best shown" (1970, p. 224) and treating them as Aristotle does the notion of "being," that is, as the referential equivocal in terms of which other things are said to have the property "as related to one central point, one definite kind of thing, and . . . not . . . by a mere ambiguity" (*Meta* 1002a30). In the most general sense, Being *is* the "act of belonging," hence it

> is hierarchical. . . . The act by which an atom is . . . is an act of being just as truly as is the act by which a man is . . . but the context of the act is in each case different. . . . Man belongs and is belonged to in a manner very greatly different from an atom. . . . As the hierarchy of being ascends, the contents of being become richer. . . . Because of the fact that being grows as it rises, and grows particularly in meaning as purpose and consciousness enter, it is possible to affirm that the meaning of being is best and most truly disclosed in the upper reaches rather than the local strata. (1970, pp. 223–24)

There are three roughly distinguishable senses of ownness here. First, there is that which applies to process when it reaches consummatory ownness in which "the meaning of being is best and most truly disclosed," and hence is the referential equivocal. Second, there is that which characterizes the process of differentiation of self and other in individualizing ownness. Third, there is that in which ownness is the structural condition for process in general.

This last and most general sense, which I will refer to as "structural ownness," is displayed in the "analogy" between the ways in which the being of the atom and that of the man both involve ownness. For man, the locus of significance is the process in which the self and other "belong to one another" as coemergents in process. In the case of the atom, we cannot speak of self and other, but only of entity and context. But there is still a *process* within which atom and field are stable modes of existence unintelligible apart from one another. The very fact that the process exists testifies to a kind of compatibility between the atom and the field that makes the process possible. It may seem odd to talk about "compatibility" here, since, given that there *are* atoms and fields, there doesn't seem to be any way they could be *in*compatible. But this is just the point. "Being" includes whatever it has taken to make possible whatever is going on. For this reason, making the structure of these processes intelligible must involve sorting out just what ranges of specific conditions—ways of belonging—are involved in any particular sort of process.

Consider the parallel issues raised by Aristotle's inquiry into the number of "principles" that must be postulated if we are to make sense of the processes of nature. He argued that there must be matter, form, and the privation of form. By this last notion he did not mean a third element, but a way of taking account of the fact that there are *conditions* for orderly process, since "our first presupposition must be that in nature nothing acts, or is acted on by, any other thing at random. . . . Nor again do things pass into the first chance thing" (*Phy* 187a32, 188b3). Something like this is what Hofstadter must mean by saying that process presupposes "ownness" between the entity and the context such that the process (which is the locus of significance) can occur. Minimally, it means that the conditions for the possible to become actual are met, and that there are *specific* processes involved.

The Interpenetration of the Three Modes of Truth

In *Truth and Art* "the question that . . . led us to . . . the concept of spiritual truth . . . was . . . whether there is a kind of truth that enables us to break the chain of our own unlimited positing of the ought" (1965, p. 130). But Hofstadter could have reached the same conclusion by arguing from a threatened regress in the truth of statement, rather than the truth of things. For neither of these polar modalities of truth are complete in themselves: each requires the other, and requires it in such a relationship as constitutes truth of spirit.

Both the process in which the self imposes its form on the other, and that where the other imposes its form on the self, call for a consummatory point where the "imposition" is not something external and arbitrary, where the governance is cogent, not coercive. The phrase "compelled by the evidence" can describe an epistemic rather than a pathological process only if the mind so compelled is free and authoritative in its acknowledgment of the evidence that ends the regress of questioning. Miller puts it this way:

> Science is much more than an enlargement of animal learning. It is also the primary and the most solid region of community. It is not a community of custom, ritual, or habit, or an anthropologi-

> cal pattern of behavior. It is a community of minds in which men exercise power over each other through their vulnerability to criticism. And it is through this freely assumed liability to error that respect is established and men becomes ends in themselves. (1978, p. 103)

When a scientist says, "This statement is warranted," she is not only offering the statement in relation to some evidence, but presenting herself as someone governed by respect for evidence, not by idiosyncrasy or some nonepistemic end. Hofstadter does not develop the implications of the fact that she can do this only insofar as she is a member of "a community of minds in which men exercise power over each other through their vulnerability to criticism." But since it is to this community that the exemplary judgment "imputes" agreement, this theme must be developed. But for the moment, let us return to the problem of the threatened regress in truth of statement.

Even that scientific statement most rigorously excluding the idiosyncrasies of the observer offers a piece of the self to the other; it presents the self as responsible for the exclusion from the judgment of all but the weight of the evidence (Hofstadter 1965, p. 84). Robert Paul Wolff gives an illustration of this in his discussion of the claim that "education without freedom is impossible" (1969, p. 102). If "education" means merely that someone can "repeat some proposition, answer questions about it, get it right on an examination . . . then knowledge has nothing to do with freedom. Or does it mean that he can give *reasons* for what he believes, that he can himself see the merits of those reasons, and that he stands ready to withhold or change his beliefs whenever he judges it *right to do so?*" (p. 102, my emphasis). The phrase "right to do so" is just the sort of epistemic obligation that can be asserted with authority only within that "community of minds in which men exercise power over each other through their vulnerability to criticism" (Miller 1978, p. 103). This community, as the condition of "cogent governance," is invoked because the statement not only presents the speaker as obedient to the command of evidence, but is addressed to a hearer who resembles the speaker in relevant respects, that is, as one whose unity of judgment can transcend any merely private point of view. In this sense, the relevant community will be constituted by whatever institutions are necessary to move from ungrounded opinion to warranted assertability.

In the narrow sense of statement truth exemplified in science, the criteria for belonging to this community are spelled out in the institutions defining the "normal" science of any day. But some form of respect for evidence governs the whole realm of statement truth, and some sort of community is presupposed for all authority. As Hofstadter says,

> In knowledge, what is well grounded in evidence has authoritative weight and effective binding power upon our rational understanding. It commands our cognitive acknowledgement, and we acknowledge that it does so rightfully. . . . [It] is not the logicians' correspondence of a proposition with a fact, but the living person's realization of the binding force of perceived or understood evidence on his belief. (1965, p. 161)

When truth of statement has reached this level, it can no longer be described in the polarized and abstract terms in which Hofstadter introduced it as "the adequation of intellect to thing" (p. 91). For this sort of "acknowledgment" of the binding power of evidence involves "intellect" (ultimately, the person) as more than passive undergoing; it involves the very particular balance of doing and undergoing that constitutes truth of spirit. Hofstadter originally presented truth of statement and truth of things as *contrasted* to each other in terms of whether self was adequated to thing or thing to self; then truth of spirit was brought in to end the regress within truth of things. But now that we can see truth of statement as equally incomplete without truth of spirit, it is clear how preliminary was his original way of talking about truth of things and statement as separable modalities, isolatable from one another and from truth of spirit.

I believe that he needed this preliminary separation in order to phrase his argument in the familiar terms of the tradition. This meant talking about knowing and acting as having separate spheres, being the outcome of distinct faculties (cf. chapter 5's discussion of the original questions as doing the same thing). But the outcome of his argument is that these initial dichotomies must be overcome if any resolution is to be reached. For in the actual dialectic within which all the forms of self and other are generated, there are no instances of doing that are not also undergoing, hence none of knowing apart from those challenges and saliences generated through the action, nor of acting apart from the cognitive structuring of the world that invites or necessitates act. These same traditional ways of talking lead us to associate truth of things with doing and the subject, contrasting it to truth of statement, which reflects, in undergoing, the form presented by object. But if doing and undergoing are defined functionally, rather than in terms of some fixed range of entities, then both terms can as easily characterize subject as object. There is no need—although there is an egotistical temptation—to assume that it is always the subject that accounts for the doing.

Nor should we assume that all doing is conscious action, or all undergoing a form of critical cognition. The ways in which self thrusts into the world include much more than those conscious acts whose self-critical evaluation poses the problems of the regress; and the thrust of the world onto the self is infinitely richer than the intrusion of intelligible form onto an ideally open intelligence.

The key point is that experience is always some equipoise, however fragmentary and tortured, between the dialectical polarities. Hence the poles are logical extremes, not actual occurrences. This is why Aristotle said that Pure Matter was only an ideal of thought: any actual thing must have a particular form. I also want to make the parallel claim (which Aristotle rejects) that form is always the form of some matter, never an essence with its own separable being. This is entailed by the claim that transactions are ontologically prior, and that subject and object, form and the receptor of form, are relative stabilities defined within process and in relation to one another. Doing and undergoing (and all their specific forms) are *ranges* of activity marked out by taking some salient macroquality of synthesis as the locus of significance.

The Demand for Ownness with All That *Is*

The salient macroquality in terms of which the ranges of doing and undergoing can be given significance is, like doing and undergoing themselves, defined functionally and formally, not in reference to some fixed set of entities or experiences. Thus "consummatory ownness" does not designate a special set of "super experiences." On the contrary, the reference is to a distinctive and salient *way* in which synthesis goes on between modes of doing and undergoing. I have tried to make this clear by using an awkward set of special locutions, referring to synthesis at the fundamental level required for process to occur at all as "structural ownness," to the levels of synthesis necessary for the emergence of individuals conscious of their alienation from the other as "individualizing ownness," and to the phase in which this alienation is overcome as "consummatory ownness." It is this final phase that serves as the referential equivocal in terms of which the others are said to be modes of ownness.

The same set of relationships holds for Hofstadter's concepts involving the three kinds of truth. Truth of spirit is the integrating notion, but it does not refer to some special revelation of truth, some breakthrough into Being itself, but to the *ways* in which truth of statement and truth of things can work together to achieve a cogent governance. As such, it points to a dynamic structure within the modes of doing and undergoing. Any level whatsoever of actual process will have at least a basic compatibility and fitness between the elements—whatever it has taken to make it possible for doing and undergoing to occur. This is the level of synthesis involved in structural ownness. It is only where synthesis has the dynamic balance in which the other is own to the self that we speak of truth of spirit.

When Hofstadter claims that the most vivid and available modes of truth of spirit are available in art, he does not presuppose that art gives us a more ultimate form of ownness than ethical or religious experience. Quite the opposite: because art deals with "semblance" and with the "appearance" of truth, it provides an introduction and anticipation of the more difficult, ultimate, and inclusive possibilities of ownness:

> Art can show us all these possibilities over and over again, as human life poses itself over and over again as a question seeking its answer. But to find the answer in reality itself, outside the restricted medium of art, is still the problem. Man exists outside the shell of the artistic medium. Therefore he must look into the world—he himself is the "must look" here—for those points at which the possibility of truth of being occur. . . . In this search he moves into the sphere of ethics and religion. Here the problem of human existence gets posed in a different way. In art, man is able to delimit the problem by restricting his universe to the universe of the symbol. His power is like that of a demiurge over the medium he can work with. He has the whole of a world in his hands because he can limit the whole to the given symbol: it is a canvas, a wall, a building, a stretch of fifteen minutes of sound and silence, a scope of six stanzas that he becomes responsible for. . . . But in life man is responsible for what he does with the whole of what *is*. (1965, pp. 211–12)

With most of this, one can hardly quarrel. If man can only come to be man as he articulates himself and a world, then clearly he must use as his "medium" all the varied aspects of his existence. The environment shaped into a world of ownness must include acts and institutions as well as paints and notes. But the fact that art is not sufficient does not mean that it is not necessary, nor does it exclude a fundamental role for art in giving an intelligible shape and image to these more inclusive responsibilities and tasks. Indeed, if the world as meaningful for man is a continual and never completed achievement, art will have the inescapable task—in religion, morals, and philosophy—of articulating the ideal that is not yet actual. In Hofstadter's repeated image of man as "an arrow pointed upward," it is art that gives an intuitive shape to the "target that is not yet," but which is urged in the exemplary act. Equally important is the fact that art can give us instances of that adverbial quality of integration we can seek to realize in more inclusive modes of activity.

Nevertheless, it is difficult to know just what Hofstadter means when he says that ethics and religion seek answers "in reality itself" rather than in the restricted "universe of the symbol." What can a good post-Kantian mean by the claim that the dialectic of spirit reaches its "most comprehensive and profound form in *religio,* in man's linking of himself to the transcendent, the ultimate, the unconditioned" (1970, p. 51)? He cannot mean that in religion we break through a veil of symbols to finally "see" (as a theoretical truth) that Reality is, in itself, simply own to man, that the Kantian Ideas are after all constitutive and not regulatory. Perhaps to say religion is ultimate means (at least) that it is aimed at the most inclusive ideal. Even if theory and practice can achieve ownness at their own level, each engages only a part of the self with an aspect of the world—the self as knower with the world as intelligible structure, the self as will with the world as consonant to will. But with the synthesis in spiritual truth, there must be *mutual* adequation. And while the truth of art can attempt to articulate all these aspects of self and world, it does so only as they can be embodied in a limited material, in "a canvas, a wall, a building, a stretch of fifteen minutes of sound and silence, a scope of six stanzas."

But the forces to be brought to expression are so primal and fundamental that, with its limited material, the work of art can be at most a "fragmentary articulation" (1970, p. 190). In painting, for example, the "artistic consciousness" is so fully concerned with "visual appearances" that "everything irrelevant to this recedes" (p. 82). The result is a combination of form and matter that inevitably achieves its unity by leaving out elements ultimately essential to full humanity. This romantic emphasis on the forces inexpressible in art grows from *Truth and Art* to *Agony and Epitaph.*[3] It becomes dominant in a later article, "The Aesthetic Impulse," where the classical aesthetic image that "shows man the reconciliation of opposites as actually accomplished" (1973a, p. 177) is portrayed as only a stage in the dialectical development of spirit toward freedom: art attempts to express "the form of ownness-relation that belongs to religion, which is ultimately an intimacy of spirit with spirit, and of all spirits within their community," but this *"cannot be adequately embodied in a sensible configuration"* (p. 180). Insofar as the tortured forms of contemporary art are an attempt to "show in sensible form the inability of sensible form to express ultimate freedom," they reveal "the spirit's tearing at the form, pulling it, distorting it, striving ever to make it express more than it can say" (p. 180).

It is not clear to me either how religion can be more adequate in this sense than art, or how philosophy is related to this claim. Perhaps philosophy represents the theoretical demand within the level of spirit, as art gives us a sense of the quality of activity that would appear in a world totally own (which, of course, no actual world can be). If so, then religion as synthesis would include a volitional component absent in philosophy. The inevitability of *some* attempt at a mode of inclusive synthesis is built into the logic of Hofstadter's position, and, if he is correct, into the vocation of consciousness. But much is unspecified about its putative shape or why, traditional usage aside, one would call it "religious." All in all, I think it less misleading to sum up these themes as "historical," and I will follow Miller in doing so.

But whatever language we use here, Hofstadter has given us an extraordinarily powerful account of the centrality of consummatory experience as the synthesis in terms of which other problems and modalities must be understood. What we must now do is assess the impact on our three original questions of this analysis in which we "conceive of a *person* as an entity whose being is *subjectivity in quest of spiritual truth or validity*" (1965, p. 134).

Chapter 9

Human Transactions as the Locus of Significance

> I personally have a vast respect for mind,
> but has nature? Mind is only a little bit of nature,
> the rest of which seems to be able to get along
> very well without it.
>
> —Sigmund Freud

What Remains to Be Done

Although I have been talking about those original questions *ad infinitum, ad nauseam,* let me repeat them one final time:

1. Is the belief that people can make responsible and autonomous choices about their lives compatible with what we know about the evolutionary and developmental processes out of which people emerge?
2. How can the choices made within biological and cultural history be nonarbitrary and compelling if there are no ahistoric norms against which they can be judged?
3. What sort of meaning can individual lives and choices have if the history within which they come to be is simply an accidental episode in an indifferent universe?

These questions are obviously as presumptuous now as they were when I first raised them; but if the argument has merit, we can now work on them without the distortions of unacceptable assumptions which gave us irreconcilable dichotomies: value is in the autonomous will *or* value is found as a fact of the world; history has a meaning apart from act *or* history is meaningless. Now, however, we can see these opposing views as moments *within* the process through which meaning emerges in time. As such, they pose questions open to productive empirical inquiry and reflective analysis.

The most radical shifts follow from taking the activity of process as the locus of significance within which internal and dependent distinctions such as "subject" and "world" are generated. This entails that no evaluative criteria for judging either can be based on the nature of either alone. As coemergent and relatively stable modes of process defined in terms of each other, they have no necessary and unavoidable properties except those presupposed by their common emergence vis-à-vis each other. Both their value and identity depend on the ways in which they are the *range* of conditions mutually necessary for some transaction salient from some point of view.

In terms of question 1, this means that human embeddedness in the processes of development and evolution is the *source* of the authority of act, rather than a threat to any legitimate sense of autonomy. Acts become responsible and nonarbitrary through their power to sustain some measure of integrity and coherence in the transactions out of which both self and other emerge, not because they correspond to norms apart from action (question 2). On the contrary, the authority and integrity of these transactions—including histories and biographies—are generated within history itself (question 3).

The basic arguments for these conclusions have already been made. Chapter 1 argued that any empirically adequate science must understand the activity of persons as "achievements" emergent from the processes through which organisms mature in environments. Only if this distinctive mode of activity is taken as the locus of significance can we determine the "mechanisms" governing this process of mutual determination between person and world.

Chapter 2 used the issues involved in an expanded and inclusive sense of health to explore the ways in which this codetermination of self and other involves the creation of measured order in those transactions through which self comes to be in the world. Levels of achieved self were seen as reflecting levels of integration. The notion of health as an integrated balance of functions anticipated later discussions of the systemic character of a person's biography.

Chapter 3 showed that Kant's insight into the codefinition of self and other can be extended to the coemergence of self and other in time. Because the self-world systems that emerge are always incomplete, always both summative and anticipatory, any judgment of them must include exemplary as well as determinant aspects.

Chapter 4 examined Miller's analysis of the processes through which finite acts generate the midworld within which both nature and man are defined. These originary acts sustain history as the context within which the past is given significance in terms of what is anticipated—hence as the matrix within which judgment and act are adjudicated in communities of criticism.

Chapter 5 argued that the continual revision of ends is inescapable because subject and object are relative stabilities determined in relation to one another, not self-subsisting substances. This process of mutual determination involves the thrust of self into the world, that of the world onto the self, and their synthesis. The formal relationships between these elements in the dialectic of process are reflected in the constitutional status of the three questions.

Chapter 6 drew out the implications of this dialectic of self-other into the principles that self and other cannot be defined independently of each other, that outcomes must be understood as both summative and prospective, and that all inquiry involves value commitment. These implications are incorporated in the notion of the locus of significance as the salient macroquality within which self and other are codefined ranges of stability.

Chapter 7 examined the ways in which inquiry into the relationships between stages of self and other is blocked by the assumption that form belongs *either* to subjects *or* to objects, rather than to the transactions within which they occur. Research built around these exhaustive and exclusive dichotomies excludes the temporal and comparative aspects necessary for empirically responsible and falsifiable inquiry.

Chapter 8 examined Hofstadter's argument that value can only be grounded in the process through which self and other are "own" to one another. The major points of the earlier arguments were restated and integrated in terms of the three levels of ownness: structural ownness is the transcendental condition of process in general, and reflects the codefinition of subject and object; individualizing ownness is the outcome of that process within which self and other come to be in distinction from one another; consummatory ownness is the quality of the dynamic relationship of this self-world system when it is vital and self-revising. The value component of inquiry is reflected in the fact that consummatory ownness is the locus made significant by the urgencies of life in the world.

In one sense, I take the arguments of these chapters as given, and will now go on to integrate their joint implications; in another sense, the power of these integrations to clarify our questions will be a further argument for them. In either case, more must be said about the ways in which vital and self-revising histories and biographies have the same dynamic structure that characterizes mutual respecting between persons. This structure provides the criteria for nonarbitrary decisions about the relative value of the elements contributing to them. These criteria must command with authority, yet be compatible with the autonomy of the self commanded. If they are to enable us to make comparative judgments between alternative lives, they must do so in a way that recognizes the inevitable and justifiable differences between lives; this is entailed by the absence of ahistoric norms, and by the originary power of acts.

Since the exemplary judgments about these acts "impute" the agreement of others, I need to say more about the role of these others in making up the critical communities within which these judgments—and these judges—can have authority. Participation in these communities is the sole foundation of the authority of self, whether as knowing, acting, or appreciating. This is another way of making Socrates' point that the unexamined life is not worth living, and Miller's point that one must not be banished from history. My aim is to show some of the ways in which the Greek emphasis on the wholeness of lives can be understood and judged in terms of formal (Kant-like) criteria specifying the conditions for these lives to be vital and self-revising.

Necessity Is in the Conditions of Process

These summaries and anticipations reemphasize that the argument revolves around those salient moments within process in which doing and undergoing reach the equipoise of consummatory ownness. Such moments of synthesis do not reflect a mere congruence between the shape of self and the shape of a world *found* as a encountered contingency; any such outcome would have a quite different significance from one that emerged as a critical equipoise of doing and undergoing. For example, imagine a world perfectly pliable to the self's every idiosyncratic wish and will: whatever imagined ecstasy of ease this might produce, there could be no encounter with measured order both found (as the truth of statement) and yet willed (as truth of things). Nor could there be such an encounter if self were infinitely pliable to every thrust of the world (ignoring for the moment the fact that in neither case would there be self or world). So the synthesis is not just a contingent *fact of fit* between self and world, but emerges in the shaping constraints governing transactions within which self and other come to be in codetermining tension. Here fit does not reflect the relationships between two fixed and permanent entities, but the moments of equipoise within process.

These considerations also rule out the possibility that the aspects of the other with which the self is at home are simply a reflection of the earlier activity of the self. Although there are obviously behaviors that mechanically impose the shape of the already existent self on the world, they are not our concern here: we are seeking that which measures and recognizes the self, not that which indulges it, or marks its already existent power. No such measures could be set by any form of the other that was just the surd and trail of the self's past will, or of the institutionalized will of some group to which the self happens to belong. Such measures could not *command.*

This is a slippery point for anyone who makes as much as I do of the activity of selves in generating the midworld. It is, after all, through the action of functioning objects such as clocks and yardsticks that even the world of nature is given the shape of space and time. Even more obvious are the ways in which institutions can reflect the shape of whatever will it was that cooperated in their genesis. As for the ways art shapes the world, Hofstadter—sounding much like Miller on the generation of the midworld—says that "an indifferent nature touched by art turns into a genuinely human world and man becomes an authentic dweller within it. . . . Technical art . . . transforms the natural environment into man's neighborhood and his home, in which he exists as in the place that belongs to him and to which he belongs" (1970, p. 55).

The obvious objection to this account of encountered ownness is that we are salting the mine: if we find what we put there, then it is hardly surprising that we are "at home" with it as "own." But what is going on here is more than the human animal being reassured by the homey traces of its earlier passing—although even this is not to be disparaged: in a world of vulnerabilities and death, any evidence of our own continuity sustains the will and spirit.

The recognition of ownness would indeed be mere vacuous rediscovery if it consisted simply in self first imposing form on the world (truth of things), then encountering this same form as a given (truth of statement). But agents do not first exist, then press their shapes upon the other: we only *are* persons insofar as we shape and revise a midworld of meanings within which both self and other can have actuality. The particular shapes of these dynamic processes are reflected in *styles*. Consequently, in art any theory that sees the truth of art in the reflection of antecedent realities will see style as essentially irrelevant, at best a kind of "packaging" of truth having nothing to do with the aesthetic dimensions of man's being. Similarly, any theory that sees art as subjectivity will find in style nothing more than idiosyncrasy. But where style is the way material is shaped into an image articulating a way of being in the world, then it shows us the specific form taken by the creative process of becoming human.

Since acting and knowing are as active in creating the midworld as is aesthetic activity, their "styles" are equally constitutive. It is in this sense that Kuhn's (1970) "normal science" is a shared and (relatively) static style of doing science that defines a way of being a knower and a known; it is in this sense that cultures and biographies are styles of being human. Here, as with art, these distinctions reflect ontology, not idiosyncrasy. But their ontological status does not derive from their being local variations of Universal Man, mere instances of a Form that differ only accidentally from one another. On the contrary, they are ontologically significant variations of the ways in which the authority of different self-world systems is worked out in different modes of transactions. This is what Hegel studied so brilliantly as *Geist,* and what cultural anthropologists such as Clifford Geertz (1983) see in the shapes of culture.

An originary style serves as the locus of significance in which *mutual* powers of self and other are marked out. Here significance is not a product of either correspondence or coherence, but of the continual balance of doing and undergoing. Hofstadter drives this point home in his discussion of the *symmetry* between self and other that must occur if the regress is to end in the cogent governance of truth of spirit: after saying that the intellect must both uncover and govern the thing, he immediately adds that "*it would be just as true* to say that it is the thing that uncovers the intellect and the thing that governs the intellect" (1965, p. 131, my emphasis).

Returning to Socrates' Question

The constitutive and irreducible nature of styles within the traditional modalities of acting, knowing, and appreciating derives from the fact that these modalities are themselves styles of the process whose outcome is the subject-world system (recall chapter 5's exploration of the interdependence of doing, undergoing, and synthesis, or chapter 8's analysis of relationship between structural, individualizing, and consummatory ownness). This is reflected in the way question 1 has shifted from a focus on the autonomy of self as moral to the authority of the self as an integrative whole for whom, as Hofstadter says, "the role of being a person is the fundamental role, the role of roles" (1970, p. 227).

Hofstadter's handling of the problem of the regress is a further illustration of the inseparability of the traditional modalities: although posed in terms of the impossibility of finding an end in action alone, the regress is ended by an "ought" that "though it includes the moral ought, is not a merely moral ought. The bad man is not bad in the moral sense alone (although he is that, too)—he is the unliving, inactual man, the man whose spirit is defective in some way, so that it is neither fully adequate as spirit nor does it adequately realize itself in the actuality of his life" (1965, p. 141). Put this crudely: if spiritual truth is the answer, then the question must have involved more than the moral problems raised by the truth of things. Thus truth of spirit characterizes all modalities, and is not some super and separable experience in which the "earlier" modalities are left behind.

With a certain amount of historical liberty ("license," if you prefer), we can describe the result of this increasing emphasis on inclusive modes of living as a return to Socrates' question, "How should one live?" As Bernard Williams points out, the reference to "one" makes explicit the assumption that the question applies in a meaningful sense to anyone, and that "it is not immediate; it is not about what I should do now, or next. It is about a manner of life" (1985, p. 4). I would add that the "anyone" in this reference must not be limited to individuals, but includes any subject pole of a transaction picked out as the sustaining condition when a meaningful mode of living is the locus of significance; the systemic character of meaning characterizing the mode of living applies as much to a family or a political entity as it does to a biological individual. This inclusive meaning should be added to Williams's further remark that "his question still does press a demand for reflection on one's life *as a whole,* from every aspect and all the way down, even if we do not place as much weight as the Greeks did on how it may end" (p. 5). Additionally (and of great import for our present concerns), "It . . . is also entirely noncommittal, and very fruitfully so, about the kinds of consideration to be applied to the question" (p. 5). In a similar vein, J. O. Urmson says of Aristotle's concept of the character one must have in order to be happy that "if we were to ask, not for what sort of a person do we feel the most moral respect, but what sort of person we should wish a child of ours to be, we shall be nearer to Aristotle's viewpoint" (1988, p. 27).

This puts the focus precisely on the systemic and integrative quality of *living* that qualifies the course of a subject-world transaction. At this level of inclusion, not only is the class of "entities" referred to wider than that of individuals, but the structures and considerations that set measured limits to actions of these entities are much wider than those governing action as ethical, where this is defined narrowly as the deliberate structuring of action so as to respect the dignity of others as manifested in action. Failure in terms of this wider task threatens the integrative core of the self, not what it is as moral in distinction from cognitive or appreciative.

Miller deals with this when he talks about what it means to fail radically in the task of becoming human. He begins with a sentence from Descartes' *Discourse on Method:* "For it occurred to me that I should find more truth in the reasonings of each individual *with reference to the affairs in which he is personally interested,* and the issues which *must presently punish him if he judged amiss,* than in those conducted by a man of letters in his study . . . that are of no practical moment, and followed by no practical consequences to himself" (1978, p. 81). Although Miller does not comment directly on the first underlined phrase ("with reference to the affairs in which he is personally interested"), I think it points to the need in this task for commitment that is "the mark of limit, and of self-identification with limit" (p. 85). This realm of *act,* where one is a participant and not a spectator, is the key to finitude as a category; here is found momentum and risk involving more than an intellectual matter, more than a moral dilemma, more than any particular purpose (back to Socrates' question). Although what is at stake are "affairs in which he is personally interested," these do not involve "an aim which is itself avoidable" (p. 82). What is threatened if we judge amiss must involve the conditions of having any authoritative and nonarbitrary aims at all (question 2).

There can be such aims only if there are unavoidable limits set by the process through which we become persons with aims in a world that both invites and threatens them. These limits can be found only in a history that "is the union of form and finitude, in the acceptance of time as a region from which one cannot escape because it is the condition of the disclosure of all necessities" (p. 83). Only here does the self's authority necessarily depend on its being an unfinished and vulnerable locus of action in an unfinished world. This entails that the identity of the self be subject to continual revision—not just as a self-critical intelligence, or as an actor open to growth, or as an aesthetic consciousness open to new forms, but as an integrated person who articulates a narrative structure within a codefining world. This is the locus of that most inclusive sense of obligation that is the ontic duty not to be banished from history.

The Problem of the Regress—One More Time

We have seen that unless the focus is kept on the *processes* that sustain and revise history, the attribution of ownness becomes a form of the pathetic fallacy in which human characteristics are projected onto the other. In this situation, when I "yield my will to the thing identifying my concept with its concept . . . [because its concept is] at root identical with my own" (Hofstadter 1965, p. 132), what I encounter is simply an aspect of the world shaped by past actions to fit my will. After all, if I can make a work of art to "fit" whatever perception happens to need, then I can try to do the same not just with "a canvas, a wall, a building, a stretch of fifteen minutes of sound and silence, a scope of six stanzas . . . but . . . with the whole of what *is*" (Hofstadter 1965, p. 212). Where this succeeds, the self is at ease in that part of the world it has put together for this purpose. What makes this move futile is that it can generate no measure that has "cogent governance" over the self: the fit of self with other is just one more fact in the world, not only contingent but contrived. We will not find compelling order here. As Miller says, only

> ontology is compulsive. The overall quality of experience or of nature is not disclosed in any event [of the will or the world] which could be some other way. It goes without saying that one does not want to make an *argument* for finality. Whatever is proved is unsure, and whatever happens is without constitutional force. To represent history as a category is to offer it as compulsive. One needs to discover how that can be done. (1978, pp. 79–80)

The compulsion of history does not lie in the weight of examples or in causal inexorability, but in the fact that it furnishes the conditions under which finite acts can be creative and critical. Only in history can we act with authority. Back to Kant: what is necessary if there are to be moral selves in a moral world is a *relationship*—articulated in action—of mutual respect between agents. I extend this argument to say that the kind of mutual respect involved in *any* community of criticism is necessary in all modalities, not simply the moral. The requirements of these communities command my action because membership in them is the inescapable condition of the very actions given measure by their limits. Ultimately, the authority of command lies in the necessary conditions for the articulation of self and other within a coherent transaction (this is the history that is the focus of question 3). Authority does not lie in an isolated self whose autonomy is defined through its very separation from the world (question 1); nor is it grounded in norms simply found in a world whose shape owes nothing to action (question 2).

This notion that necessity and command must be located in the structure of process rather than in the nature of some outcome is central to Kant's account of the exemplary aesthetic judgment:

> Were the pleasure in a given object to be the antecedent, and were the universal communicability of this pleasure to be all that the judgment of taste is meant to allow to the representation of the object, such a sequence would be self-contradictory. For a pleasure of that kind would be nothing but the feeling of mere agreeableness to the senses, and so, from its very nature would possess no more than private validity. . . . Hence it is the universal capacity for being communicated incident to the mental state in the given representation which, as the subjective condition of the judgment of taste, must be fundamental, with the pleasure in the object as its consequent. (1790, p. 217)

Only the categories presupposed for the possibility of experience itself have this necessary universality, for "nothing . . . is capable of being universally communicated but cognition and representation so far appurtenant to cognition" (p. 217). Since the aesthetic judgment does not refer the objects to concepts (in which case it would be a determinant and not an exemplary judgment), its reference to the faculties of cognition

> can be nothing else than the mental state present in the free play of imagination and understanding (so far as these are in mutual accord, as is requisite for *cognition in general*): for we are conscious that this subjective relation suitable for a cognition in general must be just as valid for every one, and consequently as

> universally communicable, as is any determinate cognition, which always rests upon that relation as its subjective condition. (p. 218)

This is dark and disputed text, but I am not concerned with its intricacies, but with a straightforward point: no matter how certain we may be of the nature of our own experience (say, of pleasure), that experience gives no grounds for saying that other people must necessarily have a similar experience (a similar pleasure) in similar circumstances. On the contrary, the "sole foundation of this universal subjective validity of the delight which we connect with the representation of the object we call beautiful" (p. 218) must rest in the relationships within process that are necessary for such experiences of pleasure to occur in *any* subjective experience for *any* person at *any* time whatsoever. For Kant, these relationships are in the "free play of imagination and understanding (so far as these are in mutual accord, as is requisite for *cognition in general).*" I propose to extend the argument by describing these relationships in terms of the free play of doing and undergoing, so far as these are in mutual accord, as is requisite for the generation of self and other to be in the relation of cogent governance.

This argument parallels Kant's account of duty as "the necessity to act out of reverence for the law," for no such necessity can be anchored in the experience of inclination, no matter how psychologically compelling: "Only something which is conjoined with my will solely as a ground and never as an effect" (1785, p. 400) can go beyond the merely pathological particularity of inclination. The effect is the particular outcome at a particular point in time, but necessity must seek the *ground* of the outcome in the inescapable structure of process:

> Thus the moral worth of an action does not depend on the result [the outcome] expected from it, and so too does not depend on any principle of action that needs to borrow its motive from this expected result. For all these results (agreeable states and even the promotion of happiness in others) could have been brought about by other causes as well, and consequently their production did not require the will of a rational being." (1785, p. 401)

Here necessity finds its roots in the process of production without which rational willing cannot come to be—hence in the dynamic structure of creative process, not in outcome.

Much the same can be said in the modality of knowing, where nothing is more instructive than the continual failure of attempts to ground necessary knowledge in some particular experience of certainty and belief, whether described as Cartesian "clarity and distinctness" or Platonic "vision" of the Forms. Kant saw that these failures called for a "Copernican Revolution" that grounded necessity in the structures of categorical synthesis, rather than in the experienced properties of its outcomes.

In all three of the traditional modalities, Kant could count on the categorical machinery of transcendental idealism to furnish fixed structures capable of giving a necessary and universal shape to process. Thus in the aesthetic judgment, the "universal subjective validity of the delight which we connect with the representation of the object we call beautiful" found its "sole foundation" in the "free play of imagination and understanding (so far as these are in mutual accord, as is requisite for *cognition in general*)" (1952, p. 218). But this solution is not open to a naturalism in which even the forms of imagination and understanding are shaped within process. From this point of view, the solution can only be sought in the structures of process that are necessary if the elements in it are to achieve "free play" in equipoise. In Hofstadter's terms, this will involve structural ownness as the fit that makes process possible in general, individualizing ownness that makes self and other possible as coordinate modes of stability, and consummatory ownness that makes the cogent governance of spiritual truth possible. In Miller's terms, such necessity and compulsion can be found only in the conditions that make it possible not to be "banished from history," but to critically revise self and other so as to sustain the authority of knowing, acting, and appreciating.

This is the jugular of the argument: being in history with authority presupposes membership in a critical community that makes it possible to move from random assertion to warranted assertability, from pleasure to beauty, from mores to morals. This fact ties together the Kantian theme of formal structure with the Platonic and Aristotelian focus on the narrative of lives within communities: the living well and doing well that is the life of happiness (*eudaemon*) is not possible without commitment to those communities of criticism within which living is shaped by aims at truth, goodness, and beauty. Such communities, however varied their material components, have formal conditions specifying the balance between self and other that enables the relevant kinds of criticism to take place. Here the "universal communicability" that grounds necessity is not found in the fixed forms of Kant's cognitive machinery, but in those relationships with others that are "universally" the conditions of "communicability," that is, of critical discourse in all modalities.

This is less horrendous than it sounds when I jam all the Kantian and process jargon together into one statement. The point reformulates the implications of the basic dialectic of process that is at the heart of the argument: the thrust of the self can achieve measured authority only in balanced equipoise against the thrust of the other; the rampant self of truth of things is incoherent and incomplete without the other that sets the limits of cogent governance. In each modality, this other constitutes the community that is a condition of the authority of the self; it is this community that is invoked in the claim of the exemplary judgment to go beyond mere subjective assertion.

The core of this is roughly recognized at the most common-sense levels: no claim to knowing or acting or judging beautiful things is more than mere personal opinion unless it takes account of where the speaker stands, and how this is related to other people's points of view; only then is what is otherwise random opinion squared with common sense. The man who announces his "verdict" on a painting, but will not show it to others or talk about it comparatively, or even point to what he finds powerful in it, is no critic, just a windbag without authority. And if he will not give his reasons for some pronouncement on nature, or let the rest of us look through his telescope, he is a quack, not a scientist. Nor is he a serious man to heed on matters of right or wrong if his judgments take no account of how the rest of us think, or of how our thinking is built into the language we share with him. Less obvious, but often present, is the notion that those who lack authority in these ways are in some sense shallower, less solid people—less "real," if I might use that overused word.

It was the Athenians who saw more clearly than anyone before or since that we can only be human animals with *logos* if we play a role in the *polis,* talk and be talked to, open ourselves up to the dismay and discovery that are only possible within action shaped by discourse (which is why their exclusion of women from the space of politics was so morally disastrous). That we *must* go on talking reflects the unfinished and open-ended nature of the processes through which we are human. Since any moment in the history of a person or a people is both summative and anticipatory, its meaning enters into experience and is changed and judged in a *social* process where it is worked out against the critical thrust of the other. In literature this is the concept of plot, that what one does enters into the world to come back and shape one's future; until then, the meaning of who one is as the doer of the act is not yet complete (Miller used to say that the only metaphysical justification for a drinking party was that the concept of plot was suspended, and that the foolish things said would not return to haunt you).

From this point of view, to act or to believe, or to love or hate, commits us to saying that something is "'*satisfactory*' . . . that the thing 'will do.' It involves a prediction; it contemplates a future in which the thing will continue to serve; it *will* do. It asserts a consequence the thing will actively institute; it will *do*" (Dewey 1929, pp. 260–61). The social practices of criticism that define different communities differ wildly in the effectiveness with which they promote or sustain vital modes of equipoise between doing and undergoing. Without doubt, the most effective of these communities of criticism is that of science, which has, as George Herbert Mead says, "incorporated the principle of revolution into institutions" (1956, p. 20). It can do this because it

> recognizes that progress is of the nature of the solution of a problem. What these problems present are inhibitions, the checking of conduct. And the solution of the problem stops this checking process, sets it free so that we can go on. The scientist is not looking ahead toward a goal and charting his movement toward that goal. . . . He is finding out why his system does not work. And the test of his solution of the difficulty is that his system starts working again, goes on. (p. 22)

The most general sort of "going on" is history itself as the process of articulating human being (Hofstadter 1965, p. 52). So Mead's advice to those who see history as falling within no further context of meaning, but who would yet make sense of progress as "an advance in which we cannot state the goal toward which we are going" (1956, p. 22), is to be in history in such a way as to build criticism and reassessment into our relations with each other.

This is why the existence of other modes of biography and history does not threaten the authority of our own chosen order and meaning: these alternative modes are the necessary other making the dialectic of vital development possible. This sort of relationship is most familiar in issues of knowing, where no claim is epistemologically persuasive unless it is worked out against the ways in which others have met the shared task of creating an intelligible world. Similarly, no proposal for a way of life is compelling apart from comparison with the ways others have met similar challenges and opportunities; only then will it speak to us with cogent governance. And no critical judgment on art compels our attention if it fails to meet the demands of sensitivity and focus articulated in the critical community. In all these cases, no act is complete apart from the invoked community of criticism whose concurrence is called for in exemplary judgment. This called-for agreement is not rooted in the "universal communicability" of a common nature, but in the common task of developing coherent narratives in response to a common vulnerability to the threat of meaninglessness, and the loss of measure and integrity. As Socrates said, the unexamined life is not worth living.

Since tolerance is grounded in the fact that we ourselves can have authority only in communities of criticism within which we recognize the authority of others, it is a duty, not a indulgence. It is not enough to claim that we feel "sympathy" for others because they are in the same human predicament, for this would be simply another fact in the world, and no possible source of cogent governance (a comment on Hume). This governance can come about only if the *differences* with others are essential aspects of the common, inescapable task of working out coherent answers to the question, "How should one live?" Given this, toleration is not just a virtue, but the condition of being fully human.

This tolerance involves valuing the narrative alternatives others explore, not just "putting up with" those one cannot persuade or compel. But it also includes the willingness to reject some of these alternatives as failures. They fail not by falling short of some ahistoric norm, but because they do not sustain and proclaim those conditions under which history as self-revising synthesis can "go on."

A critic might accept my argument that in principle the threat of regress can be met by the appeal to the formal structures of process, yet argue that no judgment on any *particular* style of being in history can be made with enough assurance to balance the risk of intolerance. She could argue that no matter how sound in general terms my account of the formal conditions of process may be, it is too vaguely vacuous to work as a criterion for particular judgments. To show how this criticism is both correct and misleading, we must turn again to the problem of schematism.

Dissolving the Problem of Schematism

Consider a Kant-style imperative as an answer to Socrates' question, something like "So live as to sustain and maximize consummatory ownness," or "So live as to maintain the distinctive vitality of history." Both express the core claim that persons are in history only if they can revise their purposes in accordance with some measure of cogent governance. For simplicity's sake, I will refer to this shared concept as the possession of "measured integrity." So we might try a "categorical" formulation such as "Act in such a way that institutionalizing your action would sustain and facilitate relations between self and other which have measured integrity." More briefly: "So live that your narrative is characterized by measured integrity." Although I have often succumbed to the delights of rewriting the categorical imperative, I will refrain here, for the point at issue is that *no* principle, however formulated, can both formulate the formal conditions under which *any* experience can have measured integrity, and also specify the material conditions for bringing it about under *these* specific conditions. As Kant said about the similar attempt to give an absolutely general definition of truth that would specify just what makes individual statements true, it presents "the ludicrous spectacle of one man milking a he-goat and the other holding a sieve underneath" (1787, B83). It is in this sense that the critic's charge that the principle cannot specify the particular is correct.

The problem of providing a principle inclusive enough to be universal and yet substantive enough to direct action (the problem of schematism) has deep roots. Aristotle complains against Plato that "it is hard . . . to see how a weaver or a carpenter will be benefited in regard to his own craft by knowing this 'good itself,' or how the man who has viewed the Idea itself will be a better doctor or general thereby" (*Nich. Ethics* 1097a8). The same criticism is made by contemporary "applied ethics" people against the irrelevance of "systematic ethics" (and made by every undergraduate in every Ethics 1000 ever taught). In discussions of Kant's imperative, it is nothing less than a scandal, for as Bradley (ignoring the role of the material maxim) pointed out about one way (not always Kant's) of reading the categorical imperative, people simply don't act for the sake of "duty" in general, but for the sake of particular relational duties, like being the father of a particular child (1927). Similarly, no one directly acts to bring about any such a barbarously named state of affairs as "measured integrity"—even if she could figure out what it meant, and believed it to be the ultimate aim of action. Instead, people act for all the inevitably conflicting reasons that philosophers have talked about for centuries: they aim to bring about specific goods and to follow certain rules. Unfortunately, there are so many goods specific to particular persons, and so many rules that contradict each other, that the only alternative to the One of an empty principle seems to be a mere Manyness in which we play show-and-tell with our intuitions.

How do these questions affect my claim that the most general answer to Socrates' question is something like "So live that your narrative is qualified by measured integrity"? Two issues must be distinguished. On the one hand, there is the claim that the regress *can* be satisfactorily ended: the endless self-evaluation of the "rampant self" seeking the truth of things can be ended with an account of the general conditions of cogent governance. This is our assurance that trying to live an examined life by choosing critically among competing ends and lives need not be futile; it rests in the fact that the claims of the self to responsibility and autonomy (question 1) are not negated by the absence of ahistoric norms (question 2), or by a history that has no meanings apart from those generated within it (question 3).

But Socrates' question calls for more than a reassurance that in principle decisions can be made nonarbitrarily. It asks for help in making particular answers to particular questions within a problematic context: *given* that one is faced, like Er, with a choice of lives,

> he will seek to understand what is the effect, for good or evil, of beauty combined with wealth or with poverty and with this or that condition of the soul, or of any combination of high or low birth, public or private station, strength or weakness, quickness of wit or slowness, and any other qualities of mind, native or acquired; until, as the outcome of all these calculations he is able to choose between the worse and the better life with reference to the constitution of the soul, calling a life worse or better according as it leads to the soul becoming more unjust or more just. (*Rep* 618)

In spite of his allegiance to transcendent unities, Plato does not deal with these problems by invoking an experience of "Justness Itself" as a moral meter-stick against which to measure individual choices (I think the doctrine of the Forms is more concerned with the problem of the regress than with making particular decisions, but this is an argument for another time). Nor does measured integrity designate a special form of experience that can serve as the touchstone for sorting out particular choices. For no matter what candidate is auditioned for this role—whether respect for dignity of others, pleasure (of whatever quality or quantity), happiness, living justly, loyalty to loyalty, or whatever—the move will fail. As Plato pointed out, the question, How should one live? does not call for an example of a good life, or even many examples, but for the universal in terms of which any example can be seen to *be* an example. To do this, "the universal" must embody a procedure, not illustrate a type.[1] Any example of a type can be dismissed as a merely contingent occurrence, the outcome of a particular history; but a principle that points to the conditions for *any* process whatsoever achieving the desirable outcome is one rooted in ontological necessity.

For all these reasons, "measured integrity" does not take a special experience of a special subject matter as an example against which other experiences measure up or fall short. It is what Miller calls a "universal," where "the universal is in the discriminating procedure, not in what is thereupon distinguished" (1982, p. 8). As such, it points to the dynamic structure of relationships between subject and other necessary if any particular process is to reach its appropriate mode of closure—however different these modes of closure are. This is the point we wrestled with in the last chapter (see "Relationships between the Levels of Ownness" in chapter 8) in Hofstadter's claim that consummatory ownness (truth of spirit) occurs in all three modes of truth, although the concrete manifestations of it will differ as the elements of synthesis differ. It is Dewey's point that "an experience" is defined adverbially in terms of the modes of unity and closure, not substantively in terms applying only to a certain subject matter.[2]

Interpreted in this way, the concept of measured integrity has built into it a lot of the methodological specification carried by the notion of locus of significance, which lays out a general procedure for understanding *any* sort of experiencing whatsoever, including those that are chaotic and mechanical, not just those that are systemic or valued. These methodological implications direct inquiry to the ranges of temporally qualified and mutually dependent entities sustaining some particular experience qualified by measured integrity; this effectively excludes the possibility that its own meaning and value is fixed and univocal, that is, independent of the elements involved in the processes that sustain it.

Substantive consequences follow from this. First, measured integrity can never form the basis for a principle that mechanically grinds out decisions when applied to a particular problem; it is not *sufficient* for decision but *procedural* in directing the particular inquiries focused within a wide range of problematic situations. Second, that inquiry is directed productively does not assure an unambiguous product, that is, a clear and decisive result. Indeed, since the exemplary and originary powers of act always have the potentiality of producing new concrete modes of equipoise between emerging forms of self and other, no judgment resting solely on a determinant rule grounded in the past can be adequate. What we *can* ask of our procedure, however, is that it direct us to a comparative examination of enabling conditions; ideally, this will give us concrete grounds for either affirming or revising our estimate of the experience taken as a locus of significance; with this knowledge of the "mechanism" presupposed by the "achievement," we will have the power to replicate and generalize it, or destroy or diminish it (cf. "Methodological Problems and Prospects" in chapter 1).

This formal notion of measured integrity can characterize anything from a brief act to a life, a culture, or history itself. What is central for our three questions is the focus on cumulative and systemic meaning, especially as it characterizes an individual person's life in ways relevant to answering Socrates' question, "How should one live?" Even here, however, narrative refers to a systemic mode of experiencing that is the outcome of processes involving subject and other; it is not limited to an agent's first-person experience of her own narrative, or parts of it (this is, again, James's point that "experience," rightly understood, is a "double barreled" word [Dewey 1958, p. 8]).

We are dealing here with *systemic* macroqualities (Weiss 1971) that maintain their salient and re-identifiable qualities in spite of variations within a range of contributing factors. This is why we can speak of the multifarious acts of "a" person as making up the biography of that person, or talk about two biographies as illustrations of the same cultural norm, or of families, professions, cultures, and eras as the relevant range of entities. Since deciding what counts as an entity at the relevant level is a matter of determining what range of relative stabilities is needed to account for the macroquality taken as the locus of significance, "real" entities will be whatever must be distinguished as a source of power to account for that macroquality. From this point of view, biological individuals have no ontological or methodological priority over families, cultures, or any other level of relative stability that has this function in process; for the same reasons, physical objects have no priority over either atoms or ecosystems.

There are significant differences between understanding experiences with such differing degrees of inclusiveness, especially if the understanding is from the first-person point of view. For example, contrast the way in which I can experience the history of my country with the way I experience the drive and sense of my own life, let alone the meaning of some specific act encompassed within a span of consciousness. Given our focus on the problem of schematism, however, it is the *similarities* that are at issue, for we are investigating how a general notion of value and a general methodology for dealing with it can be helpful in specific situations.

What we need is a formulation of the problematic decision situation that is formal and abstract enough to include both first-person decisions about action and an observer's evaluation of events; here neither agent nor observer, act nor event, can be characterized independently of the requirements of particular contexts. In other words, our formulation must not commit the fallacy of selective emphasis (Dewey 1958, p. 27) by assuming that a level of entity or span of act central for one specific problematic situation is central, *simpliciter,* for all. With this in mind, we can define a problematic situation as that point at which the summative meaning of the past must be affirmed or requalified by being absorbed into a proposed future; it is the intersection in process where the past, as outcome, is challenged because it is also originary. Where doing and undergoing do not rest either in stasis or equipoise, there is a imbalance in the ongoingness of organized experiencing that calls for reassessment: we must either assimilate the challenge into the narrative as it is, or adapt the narrative to the terms of the challenge (cf. Piaget 1965). Such challenges can be an invitation to further enhancement of the experiencing, not just a threat of diminishment or frustration.

It would be misleading, however, to think of the choice between levels of entities involved in these problematic situations as either inherently conflicting or exclusive. For example, an uncritical individualism, such as pervades contemporary American life and language (Bellah et al. 1985), takes the typical situation of choice to be that of a radically isolated individual faced with the choice of signing up for the values and criteria embodied in some group or tradition. But problematic situations are structured by people who are *already* embedded in roles and communities; even speaking a language structures the world in a socially mediated way. Generally, we just take the specific community invoked by our acts for granted—after all, being a member of it is simply who we are, whether as physicists, hard-edged expressionists, or Serbians. As such, the role of the community as internalized is part of what goes into making some situation problematic and salient, and what we do in the particular instance comes down to "This fits my narrative."

But much depends on the nature of this community: only if it is a critical community within which the continually threatened, ever vulnerable equipoise between doing and undergoing can be renegotiated, can we be agents with authority in history. Some communities, such as those of ideological nationalism, exclude criticism and reduce the agent to a mere functionary, banished from history. Someone with an addiction to imperatives could sum the situation up this way: So act as to preserve the integrity of your narrative by making it open to critical revision in the relevant community (cf. Kant's insistence on the demands of "universal communicability" [1790, p. 217]).

This convoluted description of decision within problematic situations, one that makes all the presuppositions and interrelations explicit, sounds more mysterious than it is. Consider, as an example, the ways in which Socrates answered the question, How should one live? through his extended act of defending and dying for the integrity of his life. Since he was, of all men, *at home* in the community of critical discourse in Athens, which was, of all times and places, perhaps the most intense, it is productive to take his vital life in the polis as the narrative that is the locus of significance. This disruption of the equilibrium of doing and undergoing in the polis was reflected in the problematic situation both as a threat to Socrates' life going on in its accustomed, argumentative, way, and to the larger community of rational discourse that shaped him, and that he shaped and sustained through his activities. Because of his exquisitely reflective awareness of his role in this authority-confirming community, these two levels are hardly distinguishable (indeed, he was almost disdainful of the threat to himself as a mortal person—although this may also reflect his belief in the immortality of the soul).[3]

Whether the focus is on Socrates himself, or on the more inclusive process of which he is a part, the vital and challenged way of functioning is *concrete and specific,* grounded in particular process: these are the ways in which measured integrity is displayed, and threatened, in local times and places. What is in common between these and other senses of narrative integrity is not some univocal property, but the systemic relationship between doing and undergoing that is formally stated in the motion of measured integrity. It is to the exploration of these systematic relationships that we are directed through the locus of significance.

And these are, on the whole, just what Socrates does explore in defense of his decision to die. His reasoning nicely illustrates the kind of comparative analysis called for by structuring the situation in terms of the locus of significance. Thus he makes his decision not just as a man who is an isolatable unit, but as one who is defined by his roles as Athenian, friend, and father. He is a citizen of *Athens,* with all the institutions that give citizenship within it a peculiar style and meaning; in *Crito* he addresses the Laws of Athens, not of Sparta or some abstract republic. And his citizenship in this Athens of his is the outcome of concrete histories in which he stood by the command of the generals at Potiaea, and Amphipolis, and Delium, and yet refused the command of the Thirty to put Leon from Salamis to death. Thus the problematic situation is one of this citizen with this history in a polis at this stage of its emergence, with these histories that would be threatened—contradicted—by these actions. And it is in this context, and to this community, that his exemplary act speaks out.

This community is not the Many who happen to live in the same place, or even those who speak the same language, but the critical community sharing a respect for each other as embodying reason. To *this* community he addresses arguments, not tears and theatrics. The moral issue is squarely in the public space and time into which he invites his judges, not in his own internal feelings, or theirs.

Had he fled Athens, Socrates would have both forfeited membership in the community that sustained his critical authority, and diminished the authority of this community—as, indeed, this community diminished itself by executing him. This point picks up the Kant-like notion that the emerging narrative of what it means to be Socrates-in-Athens is contradicted by some actions, fulfilled through others (chapter 3). This triggers *comparative* analysis (chapter 7) of the differing combinations of enabling factors supporting or contradicting the ongoing mode of experiencing taken as the locus of significance. So, for example, we must ask (as Plato did not) how obedience to law sustains rational discourse in Sparta or among the barbarians—not just in Athens—and not just by a man with Socrates' history, but by helots or women. If we conclude that "the same" act would have one significance for Socrates in Athens and another for a slave in Sparta, this does not show that "it's all relative," but that acts are—as I have been arguing these many pages—embedded in the sequential and simultaneous aspects of process.

In such an analysis, the emergent macroquality (such as the life of reason in the polis)—which is the locus of significance—functions as the referential equivocal for the meanings attributed to the shifting range of entities marking it out; that is, they have a shifting and equivocal significance "in reference" to it as primary. As such, it reflects the shifting balance of two tendencies. On the one hand, as a systemic property it maintains its meaning in spite of shifts in the range of contributing entities and conditions (the drive toward univocity); there are a number of actions Socrates or the Laws can take, yet sustain the critical community. On the other, its meaning as an outcome cannot be separated either from the processes out of which it emerges or into which it passes (the drive toward equivocity); the life of reasoning in the polis takes its shape from what the Laws and Socrates do.

It may seem unfair of me to illustrate these exemplary characteristics of act with an act so exemplary that we are still trying to articulate and actualize what it calls on us to do. But it takes the extraordinary (or the pathological) to set in relief the potentialities of the ordinary. But the same basic approach applies to very ordinary analysis of such troublesome notions as "a people," or an ethnic group or culture. Focusing on the salient mode of experiencing involved—rather than on isolated elements—marks out different actions and levels of entities as ranges; thus we do not start with some set of individuals identifiable apart from context (such as "the Palestinians"), but with those ways of behaving that are salient and identifiable and important in our problematic situation. These are the locus of significance (cf. chapter 7's discussions of the ways in which modes of behavior, such as "nurturing," are the beginning point for psychological understanding). If, instead, we begin with the elements, we will miss the fact that some of them, such as a common land or language, may be essential for sustaining one kind of systemic property of living, but not another; and if we neglect comparative analysis, we will miss the relationships between changing ranges and systemic properties.

Since there is, in principle, always another relevant condition or consequence, and since every outcome can be itself a beginning, there is a kind of regress in which each judgmental resting place can be challenged. But the regress is not vicious, for if the claims of the past and the openness of the future were not a continual challenge demanding resolution, there could be no authority conferred on the self for its role in that measured imposition of order that is the recognition of cogent governance. That there are no immutable ideals toward which we can direct thought, or action, or appreciation is not just some complicating and regrettable fact, but the condition of authoritative act: there can be no finite locus of authority without an awareness that whatever is chosen puts at risk the self that chooses. Action is thus an exemplary call on the future to live out this option of choice. These "live" options to be explored, even made the subject of "ultimate commitment" (Tillich 1957), include the possibility that "the same" individual act, say killing a second-trimester fetus, could be given such differing significances by being incorporated into different narratives that both, or neither, could be defended as a way of putting the self-world together. Something of this point is recognized in current discussions, which allow that the significance of pregnancy is affected by whether it is due to careless negligence or rape, and that the significance of bringing a fetus to term is affected by the conditions in the world into which it is to be born.

The critical community called upon to judge these narratives is made up of those without whom the self could not make these commitments with authority. The mode of living that is the exercise and manifestation of this critical process marks out self and community as its ranges. The variation in ranges possible here is substantial and significant, for *styles* of biography and history make a difference in what it means to be human, or to be a community. These material communities (called so in reference to Kant's material maxims) embody in different ways the principles specified in the ideal and formal notion of community as the dialectical other to any authoritative self.

I have been describing the formal and inclusive community as built on respect for these different ways of meeting the common threat of meaninglessness. But it can equally be described as aimed toward the common end of measured integrity. This systems property can be achieved in numerous ways; indeed, to explicate its meaning is to specify the formal conditions for generating compatible differences. Let me put this (as usual) in reference to Kant: I have been reading him as arguing that respecting others as embodiments of the moral law is categorically imperative for persons because only through that activity can they be selves whose authority embodies that of the moral law. I am now arguing that respecting others as embodiments of *differing* ways of working out the common human task is categorically imperative for persons because only through that activity can they be selves whose authority is that of one who shares in the *generation* of the structure of human meaning. Obviously, that "generation" means that finite action is originary, not that it is capricious. For the order generated is that of the necessary and common conditions presupposed by differing ways of working out a life characterized by measured integrity. We can identify and compare these "differing" ways only within a matrix that includes all of them; this necessitates an absolutely general notion of the history of persons as constituting a single community (compare Kant on the Kingdom of Ends [1785, p. 433]).[4]

The Emergence of Meaning in Time

As the general narrative of persons in a single community, history occurs in the context of nature, and this in two senses. The most ultimate is the notion of whatever "is" that makes experiencing possible, and enables history to happen. This must remain essentially negative (cf. Kant's negative sense of noumena, 1787, A252–B309), since any positive characterization would concern the ways in which "it" (neither singular nor plural is appropriate here) is made actual in action, knowing, or appreciating. This positive sense of nature as the actualized other is the second sense of nature, and is what we usually have in mind when we ask how nature limits and supports history.

For example, deep ecologists see nature as intrinsically valuable, but are opposed by those who (shallowly?) hold that any value in nature is extrinsic, the mere reflection of human needs and acts (Naess 1973).

This shared assumption that value must be *either* in self *or* in the world is what links the question, "Does nature have value because we approve of it, or do we approve of it because it has value?" with that Socrates asked Euthyphro, "Is what is holy holy because the gods approve it, or do they approve of it because it is holy?" (*Euth* 12.10). Thus the deep ecologists think that if intrinsic value is limited to the self, then we can have moral grounds for limiting our (mis)treatment of nature only if nature is a self. The attempts to justify some such "extended" notion of selfhood have been rhetorically exciting and politically activating, but epistemically implausible. And unnecessary.

It is more productive to take as the locus of significance the vital and choiceworthy modes of experiencing marking out persons and the nonhuman environment as sustaining elements. Then nature appears as the relative stability marked out as the range of conditions necessary to account for all that happens, whether in knowing, acting, or appreciating—or in history as a whole. But even in the limited sense of nature as "wilderness," which especially concerns ecologists, there are transactions that are qualified by consummatory ownness *only when* nature as wilderness is recognized as the necessary range. The limits this sets to the ways in which nature can be treated are more inclusive than those set by relationships of obligation to other selves—for nature is not a self, or composed of selves. But these limits establish obligations for which we can give arguments, not merely pleas.

To go beyond this and ask for some account of nature as having intrinsic value "in itself" is dialectical precisely in Kant's sense of attributing to the world apart from experiencing just those structures that are a condition of experiencing. The longing, deep-rooted and perhaps ineradicable, is for a universe in which history is not "simply an accidental episode in an indifferent universe" (question 3). But, as I have tried to show, not only can human lives have meaning without this, but the acts that generate this meaning have authority only because history is a context of radical challenge that *cannot* derive its norms from the universe.

This does not mean that the indifference of the universe to history is a matter of indifference to us. On the contrary, nature's indifference not only permits and enables meaning-permeated human transactions, but conditions and destroys them. "Nature" is not only wine-dark seas and powder snow under the aspen trees, but HIV and wasteful death and endless "holes of oblivion" (Arendt 1963, p. 232) that fragment living, rather than sustain and enrich it in fruitful tension. It is the death of persons and of history itself. Any adequate notion of it must include not only its power to be that realm in which we are "at home" in knowing, acting, and appreciating, but to be that other which is alien precisely in the sense that we can never be "own" with it.

Nature is also "other" in the more ultimate sense that its limited and temporary tolerance of history means that history as exemplary act calls for a response that cannot be: the end of history is an outcome that begins no further process. As such, it lacks whatever further meanings it invokes, yet cannot even try to actualize. Problematic situations are defined by challenges calling for action to reaffirm or change the meaning of emerging narratives, but the end of history does not call for action: it excludes its possibility. It is not the revision of meaning, but its end. Hence the end of history does not define a problematic situation (a real problem is one you can do something about). Nonetheless, it sets the ultimate context within which all problematic situations do occur, that is, in an "indifferent universe" where act must generate its own authority and limits.

In this universe our acts may fail to "come off," plunging us into those "holes of oblivion" so graphically described by Hannah Arendt (chapter 3). It may, *but not always*. That our moral luck (Williams 1981) is sometimes disastrous is hardly surprising, given that the biggest piece of moral luck is that there happens to be consciousness and history and morality at all. For as Miller says, "The idea of a moral universe is not the same as that of a benevolent universe. It is rather the idea of the primary influence of the forces which discover the meaning of good and evil, and therefore include both" (1978, p. 93).

What is evil is not the fact that history will bc absorbed into the universe without a trace, but those actions that deliberately set out to destroy the conditions of history. What I mean is this: evil is an emergent organizing principle that, like morality, involves the recognition of the dignity of others. For while immorality lies in ignoring the dignity of the other for the sake of some nonmoral purpose, the evil person acknowledges the dignity of the other, and acts to destroy it. She blocks or diminishes it not for the sake of something else, but for its own sake. This is the deliberate destruction of the other whose recognition sets the limits that place us in history.

Torturing someone for a political motive, or even out of pleasure in her pain, is despicable, vicious, immoral, but not evil. The pure motive to evil can be isolated only in the way Kant isolated the good will, that is, through instances in which this motive is set off *against* all others, even though most (probably all) actual instances involve a combination of motives, an overdetermination of act. So, for example, to torture someone not only without practical aim or even sadistic pleasure, but against all desires and happiness, would be simply evil.[5]

There is at first something indecently egocentric in this claim that even good and evil emerge only within time and experience; it seems incongruent with a just appreciation of our total insignificance in this indifferent universe. This would be true if value were a property of the subject projected into nature, but value qualifies the transaction, not either subject or nature apart from it. Nor is it the case that even our own experiences are only of our own contributions; indeed some experiences are of what used to be called "the sublime," that is, of nature as overwhelming our merely finite capacities; this is nature as the source of unlimited powers, of which our actualization is but an infinitesimal fragment. The fact that it is *our* experience does not exclude the possibility that it can be an experience of our own insignificance in the face of the universe that allows us to happen for an instant.

In this regard, I think of the way my colleague, Bertram Morris, died: when he was so weak that he could not sit up, and had to sip his Scotch through a bent straw, lying there arguing politics and metaphysics, I asked him if he were afraid to die. He said, "Matter comes together, matter goes apart"; but he said this not in despair, but in recognition of the context in which he kept on loving his family, and teaching, and even writing one final blast at some particularly annoying piece of nonsense in the *New Republic*. I do not think that his authority is diminished by the fact that our coming into being and passing away in the moment of history does not matter, except to him, and to us. Nor can I imagine a richer exemplification of the dignity of persons than those who fought in the maw of Holocaust, or in the way a generation of gay men have put together lives in the face not only of death, but of indifference and hostility. So my son-in-law, Eddie Stone, broke through AIDS to direct brilliant theater, be outrageous in Nantucket, and love my son.

Surely, a nature that lets these lives happen meets even Kant's conditions: "Nature must consequently also be capable of being regarded in such a way that in the conformity to law of its form it *at least harmonizes* with the possibility of the ends to be effectuated in it according to the laws of freedom" (1790, p. 176, my emphasis). That the ends we effect are our own, not those of the universe, may not be enough to make us fully at home in it. But it will have to do. For unless we both recognize and transform these limits, our aims will be uninformed by intelligence and our existence untouched by the ideal. Methodologically and ontologically, Kant without Darwin is empty, Darwin without Kant is blind.

Notes

Introduction

1. To avoid the awkward expressions "his and/or her," I use the feminine pronoun except where I am commenting on a text that uses the male.

Chapter 1

1. At this stage of the argument, I am taking Kant's (and our culture's) notion of dignity as sufficiently clear to work within the analysis of the ethical as an emergent level. Later, I will explicate the notion of dignity in process terms, showing how it functions as the "locus of significance" for investigation (see "Irreducible Outcomes in Development," in chapter 7). What I am not assuming is that dignity is a fixed and univocal property that defines the species.
2. The notion that fundamental rules must be procedural rather than fixed and substantive becomes increasingly important as the argument turns toward the centrality of history and the emergence of new modes of being human. For only procedural and formal specifications are able to take account of new forms. Ultimately, these rules find their necessity and justification in specifying the conditions of history and biography, that is, of emerging and critical human process. See "Necessity Is in the Conditions of Process" in chapter 9.
3. Kant also suggests that the purpose of a thing may suggest new mechanistic explanations: "For where ends are thought as the sources of the possibilities of certain things, means have also to be supposed. Now the law of the efficient causality of a means, considered *in its own right,* requires nothing that presupposes an end, and, consequently may be both mechanical and yet a subordinate cause of designed effects" (1790, p. 414).
4. The notion that participating in a critical community—a "community of

concurrence"—is essential for the emergence of these distinctly human activities is developed in chapters 3 and 9.

5. The methodological questions raised by the claim that some achievement or stage is emergent are examined in chapter 7 in relationship to issues of nature-nurture and child development. The criterion for a separate stage is easy to state, but sometimes difficult to apply: one looks for continuity of macrobehavior along with variations in the elements and conditions of the behavior (Weiss 1973, pp. 40–42).

Chapter 2

1. The justification for this line of argument is given in a passage that Kant takes to be a short but sufficient deduction of the possibility of theoretical knowledge by the epistemic self (1781, Axvii). I insert (rough) moral parallels in brackets:

> There are only two possible ways in which synthetic representations [practical principles of the will] and their objects [other persons] can establish connection, obtain necessary relation to one another, and, as it were, meet one another. Either the object [the other person] alone must make the representation [reason as practical] possible, or the representation [reason as practical] must make the object [the other person's dignity as commanding respect] possible. In the former case, this relation is only empirical [ethically heteronomous] and the representation [the rational practical will] is never possible a priori [other than as determined by the contingencies of desire and circumstance]. This is true of appearances [prudential relationships to other individuals] as regards that element in them which belongs to sensation [desire]. In the latter case, representation [synthesis of the manifold of desires] in itself does not produce its object [the other person] in so far as existence is concerned, for we are not here speaking of its causality by means of the will. None the less the representation [reason as practical] is a priori determinant of the object [the individual as a person], if it be the case that only through the representation [the organizing principle of the will] is it possible to know [act rationally toward] anything as an object [anyone as a person].

Interpreted in terms of moral experience, this says that the dignity of the other is not the "discovery" of an external state of affairs demanding my capitulation to its power; nor is it a projection of myself, an arbitrary baptism of the other's agency as a gratuitous gift of my own already existent agency. It is, rather, the simultaneous coming into being of a self with the kind of world through which its agency is defined. Only then can the imperatives found in this world be the self's own, not imposed by violence from without.

Chapter 3

1. That the relevant community of concurrence is in some measure produced by the act parallels in a suggestive way yet another Kantian distinction: "For *estimating* beautiful objects, as such, what is required is *taste;* but for fine art, i.e., the *production* of such objects, one needs *genius*" (1790, p. 311). The productions of genius are not in accordance with an existent rule, else they would be mere imitation: "Rather must the rule be gathered from the performance, i.e., from the product, which others may use to put their own talent to the test, so as to serve as a model, not for *imitation,* but for *following*" (p. 309). What I am arguing is that the relevant sense of "following" in the moral life involves the communities of criticism projected by the act. What Kant says here also suggests an alternative to positivistic characterizations of art as whatever is called such by the linguistically dominant group. For the notion of the work of genius as an exemplary production that is "followed" but not "imitated" gives the basis for the projection of a community of criticism within which evaluation, and not just description, is possible.

Chapter 4

1. Again this parallels Dewey: "It is a plausible prediction that if there were an interdict placed for a generation upon the use of mind, matter, consciousness as nouns, and we were obliged to employ adjectives, and adverbs, conscious and consciously, mental and mentally, material and physically, we should find many of our problems much simplified" (1958, p. 75).

Chapter 6

1. I want to express my deep gratitude to my friend and colleague, Professor Eugene Gollin, Department of Psychology of the University of Colorado, for his many years of patient instruction in the developmental sciences. The notion of "locus of significance," which is the core of this chapter, was hammered out over many sessions with him; the chapter itself is based on the article "On the Uses of the Concept of Normality in Developmental Biology and Psychology," for which he is first author.
2. The definition of significance as it applies within biology and psychology is conditioned by the notion of "constraints," which include: (1) those imposed by biology and developmental history; (2) those that inhere in resource availability; and (3) those that derive from developmental ordering principles (Alberch 1983; Lindenmayer 1982). The ordering principles govern the transactive character of organism-niche relationships in that they express physico-chemical laws. A major task of research is the specification of the developmental ordering rules that characterize morphogenesis and behavioral development within the framework of these constraining factors.

Chapter 7

1. This point is illustrated both in the emergent levels of "health" looked at in chapter 2 and in chapter 9's exploration of the ways in which "ownness" characterizes all the modes of human articulation. How terms shift in meaning in these ways is laid out in Aristotle's analysis of homonymy, *Categories* 1a1–4, 6–7.
2. It is essential to emphasize at every stage of this argument that the disagreement between Johnston and the Lorenzian approach is *not* over whether "genes contribute to the process of development: . . . disagreement arises over whether the genes can be said to specify *directly* any aspect of the behavioral phenotype" (Johnston 1988, p. 623, my emphasis).
3. It is worth substituting reference to moral concepts for arousal states in this passage: "a morally qualified act is something typically attributed to the agent, but moral action is not understandable aside from a morally relevant situation, and the morally relevant aspects of a context cannot be identified independently of knowledge about the individual. When moral agents encounter a situation that affects them as morally relevant, their moral attitude changes. The moral quality of the action is actually a characteristic of the agent in the world."

Chapter 8

1. It is not clear exactly which theorists, at which times, Hofstadter takes to exemplify each stage of the dialectic, although assignments to broad periods are obvious. So it seems safest—and quite adequate—to take the stages as exemplifying, at least but not at most, a *conceptual* dialectic.
2. The fact that Hofstadter talks about statements as the unit of analysis does not mean that he rejects the notion that statements are only intelligible within some stated or implied context. Quite the contrary. Compare *Agony and Epitaph,* Preface XII (1970, pp. 206–8, 247), and *Truth and Art* (1965, p. 90). See also his discussion of philosophy as—by definition—the search for truth, in "Philosophy as the Confession That Being Is Communion" (1971).
3. See Hofstadter 1965, pp. 173, 209–10, 156, 167, 172, 190, 200–201; and 1970, pp. 35, 53–54, 67, 194, 200, 214, 223.

Chapter 9

1. See, for example, the reasoning in such passages as this:

> The just man does not allow the several elements in his soul to usurp one another's functions; he is indeed one who sets his house in order, by self-mastery and discipline coming to be at peace with himself, and bringing into tune those three parts. . . . Only when he has linked these parts together in a well-tempered harmony and has made himself one man instead of many, will he be ready to go about whatever he may have to do. . . . In all these fields when he speaks of just and honorable conduct, he will mean the behavior that helps produce and preserve this habit of mind; and by wisdom he will mean the knowledge which presides over such conduct. (*Rep* 4.443)

2. The contrast between experience of a special kind (marked by a verb) and a special kind of experiencing (marked by an adverb) is a familiar one; thus Aristotle waivers on whether the life of happiness is the special experience of contemplation, or the doing well of all the other things we do (a One apart from the Many actions, or a One among the Many). But good twentieth-century Aristotelians like John Dewey must take form as the adverbial mode of process rather than something simply disclosed in process. If it were the latter, we could always dismiss it as a contingent outcome; but if

it is a condition of successful knowing and appreciating and acting, then it is the ontological source of necessity. This is the point of Kant's argument that the "sole foundation of this universal subjective validity of the delight which we connect with the representation of the object we call beautiful" (1790, p. 218) must be in the universal conditions of cognition, not in a particular sort of pleasure taken as an outcome.

3. This sense of the identity between inner and cultural selves is described by Erik Erikson as "a *subjective sense* of an *invigorating sameness* and *continuity*" (1968, p. 19); to this he adds James's description of it as "an element of active tension, of holding my own, as it were, and trusting outward things to perform their part so as to make it a full harmony, but without any *guaranty* that they will. Make it a guaranty . . . and the attitude immediately becomes to consciousness stagnant and stingless."

   Much in the same line is Freud's phrase, "die Heimlichkeit der innerren Konstruktion," which Erikson translates as the "safe privacy of a common mental construction" (1968, p. 21).

4. The dynamic structure of this historical process is beautifully captured in Hofstadter's image of man as the arrow. The core of his argument is that the work of putting together an inclusive narrative is ontologically generative: "The activity that is constitutive of man's being is the continuous activity of distinguishing out and relating together the self and its world. This activity is genuinely creative, for in it man's world and self are originated and maintained, not found already finished" (1965, p. 38). *Were* it finished, there would be a target to aim at, and determinant judgments to be made about hitting it. But the point is *"to arrive at a target which is not yet there"* (1965, p. 35), for "Man is an arrow pointed upward" (p. 134; see also pp. 99, 173, 209–10).

   This image (in which I think Hofstadter is invoking Heracleitus) brings the mutual dynamic of intentional aim and intended target to a concrete intuition. The point I have been hammering at about the community of confirmation means that one must think of this aiming as cooperative. This does not mean that man as the arrow is "aimed" by another (although that does occur when action is not authentic), but that he is an aiming *with* another in an act that defines both self and other in relation to this target. So to speak of "community" here refers to the fact that the relationships between the persons involved are part of who they are, not just external relationships they can pass in and out of and remain the same.

   The phrase "not yet there" does not mean "not now but later," for the arrow will never reach a target but will be consumed in flight. With the end of history, it will simultaneously disappear into process along with its dialectical other, the target. Its "aim" is only to come to be as coherent aiming, measured against the target that will never "be" as a fixed terminus. So the "yet" does not refer to a fixed point that is yet to be.

"Yet" as a modality of time is itself generated in the process. It is the measure of potentiality to actuality in the act that summons both the target and the aim into being. Thus it is a mark of the incompleteness of this process within which man summons the target and the target summons man. This process gives measure to the intrinsically unmeasured will (although "will" is too narrow) and to the potential infinity of possible aims. It eliminates some aims as not sustainable. Thus the target is that which must be if the arrow is to come to be as a measured aim. The target is the arrow's *own* target; it is *man's* target, or even *this* man's target, not by assignment from without, but only through the recognition of the target by the human arrow.

If the end of the regress is the imposition of measure (i.e., the positing of a target that makes possible authoritative revision of the target), then it involves whatever conditions are necessary to permit the critical correction of aiming to go on. This process is not the arresting of change, but the ordering of change that is the articulation of intelligibility within process. At the human level, this is the notion of a coherent narrative that answers the question, How should one live?

So even the image of the "aim of the arrow" is too static. We need instead the flying of the arrow, indeed, that aspect of flight that displays the continuing relationships between the aim and the target. It is the ordering of flight—the narrative as the systemic property of process—which is the locus of significance in terms of which both the self and the target are marked out.

In the midst of all this talk of the unfinished and the not yet, it is essential to keep in mind that participating in this process can be immediately consummatory and fulfilling. Without this there would be no criteria for revision. These aspects of process are most intensely accessible in art, most inclusive in living itself.

This is all quite idealized (as well as metaphorical) talk, particularly if one thinks of the all-too-ordinary situation of people locked into Kierkegaardian "shut-upness," too unauthentic even for dread, let alone its vital overcoming (1844, pp. 110ff.). To the extent that my acts aim at accidental images unchosen and unrevised, I lack the authority of being a self-measuring aim. To the extent that every change in the world redefines my target, my self, I lack the authority of being a durational and stable self: "I am nothing if each situation can construct me anew" (Becker 1964, p.

69). Clearly, to be finite and fallible is to fall short of this ideal of critical self-revision, even at the human best; and most of us are defined more by our failures than by our moments of critical authority. But failures can only be recognized as such in reference to the self-correcting flight that is characterized by measured integrity.

5. If evil motives are as ultimate and equally "for themselves" as good, then we must ask if the evil person's principle can equally so order her self/world to constitute a realm of evil space and time within which her actions can have authority. I think the short answer is that the evil person has no other whose existence sets limits to her will, and hence brings about the end of the regress in cogent governance. Particular evil actions have limits only insofar as dignity exists as a target, that is, as evil is unsuccessful. But, as Hofstadter says, "Nothing intrinsic to the vanity of evil sets a bound to its desire. As this singular, particular ego, it wishes to be absolute" (1973b, p. 19). As such, it is unlimited and unmeasured, hence self-destructive. Hofstadter's *Lectures on Evil* (1973b) are the most provocative reflections I have found on these dark matters.

Bibliography

Aiken, Henry David (1962). *Reason and Conduct.* New York: Alfred Knopf.

Alberch, P. (1983). "Mapping Genes to Phenotypes, or the Rules of the Game That Generate Form," in *Evolution* 37:861–63.

Arendt, Hannah (1963). *Eichmann in Jerusalem: A Report on the Banality of Evil.* New York: Viking Press.

Aristotle (1941). *The Basic Works,* ed. Richard McKeon. New York: Random House.

Baltes, P. B. (1979). "Life-Span Developmental Psychology: Some Converging Observations on History and Theory," in *Life-Span Development and Behavior,* ed. P. B. Baltes and O. G. Brim. Vol. 2. New York: Academic Press.

_____ (1983). "Life-Span Developmental Psychology: Observations on History and Theory Revisited," in *Developmental Psychology: Historical and Philosophical Perspectives,* ed. R. M. Lerner. Hillsdale, N.J.: Erlbaum.

_____ (1987). "Theoretical Propositions of Life-Span Developmental Psychology," in *Developmental Psychology* 23:611–26.

Baxley, G. B., and J. M. LeBlanc (1976). "The Hyperactive Child: Characteristics, Treatment, and Evaluation of Research Design," in *Advances in Child Development and Behavior,* ed. H. W. Reese and L. P. Lipsitt. Vol. 11. New York: Academic Press.

Becker, Ernest (1964). *Revolution in Psychiatry.* New York: The Free Press.

_____ (1973). *The Denial of Death.* New York: The Free Press.

_____ (1975). *Escape From Evil.* New York: The Free Press.

Bellah, Robert N., Richard Madsden, William M. Sullivan, Ann Swindler, and Steven M. Tipton (1985). *Habits of the Heart: Individualism and Commitment in American Life.* New York: Harper and Row.

Bernstein, Richard (1986). *Philosophical Profiles: Essays in a Pragmatic Mode.* Philadelphia: University of Pennsylvania Press.

Blanck, Gertrude, and Rubin Blanck (1974). *Ego-Psychology: Theory and Practice*. New York: Columbia University Press.

Bolger, Niall, Avshalom Caspi, Geraldine Downey, and Martha Moorehouse (1988). *Persons in Context: Developmental Processes*. Cambridge: Cambridge University Press.

Borger, Robert, and Frank Cioffi, eds. (1970). *Explanation in the Behavioral Sciences*. Cambridge: Cambridge University Press.

Bowlby, John (1988). *A Secure Base*. New York: Basic Books.

Bradley, F. H. (1927). "Why Should I Be Moral?" in *Ethical Studies*. Oxford: Oxford University Press.

Brody, Howard, and David S. Sobel (1979). "A Systems View of Health and Disease," in *Ways of Health,* ed. Brody and Sobel. New York: Harcourt, Brace, Jovanovich.

Buber, Martin (1970). *I and Thou*. New York: Scribners.

Cassirer, Ernst (1953–57). *Philosophy of Symbolic Forms,* trans. Ralph Mannheim. New Haven: Yale University Press.

Chauvin, Remy (1977). *Ethology: The Biological Study of Animal Behavior.* New York: International Universities Press.

Cirillo, Leonard, Bernard Kaplan, and Seymore Wapner, eds. (1989). *Emotions in Ideal Human Development*. Hillsdale, N.J.: Erlbaum.

Cooper, R. M., and J. P. Zubek (1958). "Effects of Enriched and Restricted Learning Environments on Bright and Dull Rats," in *Canadian Journal of Psychology* 12:159–64.

Crawford, Donald W. (1974). *Kant's Aesthetic Theory.* Madison: University of Wisconsin Press.

Danto, A. C. (1985). *Narration and Knowledge*. New York: Columbia University Press.

Dasen, P. R. (1972). "Cross-Cultural Piagetian Research," in *Journal of Cross-Cultural Psychology* 3:23–40.

Dewey, John (1929). *Quest for Certainty*. New York: Minton Balch and Co.

_____ (1934a). *Art as Experience*. New York: Minton Balch and Co.

_____ (1934b). *A Common Faith*. New Haven: Yale University Press.

_____ (1958). *Experience and Nature*. New York: Dover.

Dewey, John, and Arthur Bentley (1949). *Knowing and the Known*. Boston: Beacon Press.

Diefenbeck, James E. (1990). "Act and Necessity in the Philosophy of John William Miller," in *The Philosophy of John William Miller,* ed. Joseph Fell. Lewisburg, Pa.: Bucknell University Press.

Dubos, Rene (1979). "Medicine Evolving," in *Ways of Health,* ed. Howard Brody and David S. Sobel. New York: Harcourt, Brace, Jovanovich.

Elkind, David (1967). "Egocentrism in Adolescence," in *Child Development* 38:1024–34.

Erikson, Erik (1968). *Identity, Youth, and Crisis*. New York: Norton.

Fischer, K. W., and L. E. Silvern (1985). "Stages and Individual Differences in Cognitive Development," in *American Review of Psychology* 36:613–48.

Fisher, W. R. (1984). "Narration as a Human Communication Paradigm: The Case of Public Moral Argument," in *Communication Monographs* 52:347–67.

Frankena, William (1951). "Moral Philosophy at Mid-Century," in *Philosophical Review* 60 (1):45–55.

Frankl, Victor (1963). *Man's Search for Meaning*. New York: Pocket Books.

Geertz, Clifford (1983). *Local Knowledge*. New York: Basic Books.

Gellantly, Angus, Don Rogers, and John A. Sloboda, eds. (1989). *Cognition and Social Worlds*. New York: Oxford University Press.

Gibson, J. J. (1966). *The Senses Considered as Perceptual Systems*. Boston: Houghton-Mifflin.

Goldstein, Kurt (1939). *The Organism: A Holistic Approach to Biology Derived from Pathological Data in Man*. New York: American Book Co.

Gollin, Eugene, ed. (1981). *Developmental Plasticity: Behavior and Biological Aspects of Variations in Development*. New York: Academic Press.

_____, ed. (1984a). *Malformations of Development: Biological and Psychological Sources and Consequences*. New York: Academic Press.

_____ (1984b). "Early Experiences and Developmental Plasticity," in *Annals of Child Development* 1:239–61.

_____, ed. (1985). *Comparative Development and Adaptive Skills: Evolutionary Implications*. Hillsdale, N.J.: Erlbaum.

Gollin, Eugene, and Gary Stahl (1985). "'Hyperactivity': A Misleading Label," in *Outlook* 56:12–18.

Gollin, Eugene, Gary Stahl, and Elyse Morgan (1989). "On the Uses of the Concept of Normality in Developmental Biology and Psychology," in *Advances in Child Development and Behavior,* ed. Hayne W. Reese, pp. 49–71, Vol. 21. San Diego: Academic Press.

Gottlieb, Gilbert. (1985). "On Discovering Significant Acoustic Dimensions of Auditory Stimulation for Infants," in *Measurement of Audition and Vision in the First Year of Life: A Methodological Overview*. Norwood, N.J.: Ablex.

Goulet, L. R., and Paul B. Baltes (1970). *Life-Span Developmental Psychology: Research and Theory*. New York: Academic Press.

Guyer, Paul (1979). *Kant and the Claims of Taste*. Cambridge: Harvard University Press.

Hartmann, Heinz (1958). *Ego Psychology and the Problem of Adaptation*. New York: International Universities Press.

Havighurst, R. J. (1973). "History of Developmental Psychology: Socialization and Personality Development Through the Life-Span," in *Life-Span Developmental Psychology: Personality and Socialization,* ed. P. B. Baltes and K. W. Schaie. New York: Academic Press.

Hofstadter, Albert (1965). *Truth and Art*. New York: Columbia University Press.

_____ (1970). *Agony and Epitaph*. New York: Columbia University Press.

_____ (1971). "Philosophy as the Confession that Being Is Communion," in *Monist* 55:255–74.

_____ (1973a). "The Aesthetic Impulse," in *Journal of Aesthetics and Art Criticism* 32 (2):171–81.

_____ (1973b). *Reflections on Evil* (The Lindley Lecture). Manhattan: Department of Philosophy, University of Kansas.

Jahoda, Gustav, and I. M. Lewis, eds. (1988). *Acquiring Culture: Cross Cultural Studies in Child Development*. New York: Croom Helm.

Johnston, Timothy D. (1987). "The Persistence of Dichotomies in the Study of Behavioral Development," in *Developmental Review* 7:149–82.

_____ (1988). "Developmental Explanation and the Ontogeny of Birdsong: Nature/Nurture Redux," in *Brain and Behavioral Sciences* 11:617–63.

Kant, Immanuel (1781, 1787). *Immanuel Kant's Critique of Pure Reason,* trans. N. Kemp Smith from *Kritik der reinen Vernuft.* London: Macmillan, 1929.

_____ (1785). *The Moral Law,* trans. H. J. Paton from *Grundlegung zur Metaphysik der Sitten*. London: Hutchinson, 1948. Pagination is in accord with Royal Prussian Academy edition.

_____ (1788). *Kant's Critique of Practical Reason,* trans. Lewis White Beck from *Kritik der praktischen Vernuft*. New York: Bobbs-Merrill, 1956. Pagination is in accord with Royal Prussian Academy edition.

_____ (1790). *Kant's Critique of Judgment,* trans. J. C. Meredith from *Kritik der Urtheilskraft*. Oxford: Clarendon Press, 1952. Pagination is in accord with Royal Prussian Academy edition.

Kierkegaard, S. A. (1844). *The Concept of Dread,* trans. Walter Lowrie. Princeton, N.J.: Princeton University Press, 1957.

Kohlberg, Lawrence (1969). "Stage and Sequence: The Cognitive-Developmental Approach to Socialization," in *Handbook of Socialization Theory and Research,* ed. David A. Goslin. Chicago: Rand-McNally.

_____ (1971). "From Is to Ought: How to Commit the Naturalistic Fallacy and Get Away with It in the Study of Moral Development," in *Cognitive Development and Epistemology,* ed. Theodore Mischel. New York: Academic Press.

_____ (1981). *The Philosophy of Moral Development: Moral Stages and the Idea of Justice*. San Francisco: Harper and Row.

Kuhn, Thomas (1970). *The Structure of Scientific Revolutions*. Chicago: University of Chicago Press.

Lehrman, Daniel (1953). "A Critique of Konrad Lorenz's Theory of Instinctive Behavior," in *Quarterly Review of Biology* 28:337–63.

_____ (1970). "Semantic and Conceptual Issues in the Nature-Nurture Problem," in *Development and Evolution in Behavior,* ed. L. R. Aronson, E. Tobach, D. S. Lehrman, and J. S. Rosenblatt. Chicago: W. H. Freeman.

Lerner, Richard M., ed. (1983a). *Developmental Psychology*. Hillsdale, N.J.: Erlbaum.

_____ (1983b). "The History of Philosophy and the Philosophy of History in Developmental Psychology: A View of the Issues," in *Developmental Psychology: Historical and Philosophical Perspectives*. Hillsdale, N.J.: Erlbaum.

Levi, Primo (1969). *Survival in Auschwitz: The Nazi Assault on Humanity*. New York: Collier.

Lewontin, R. C. (1974). "The Analysis of Variance and the Analysis of Causes," in *American Journal of Human Genetics* 26:400–411.

Lindenmayer, Aristid. (1982). "Developmental Algorithms: Lineage Versus Interactive Control Mechanisms," in *Developmental Order: Its Origin and Regulation,* ed. S. Subtelny and P. B. Green. New York: Alan R. Liss.

Lorenz, K. Z. (1937a). "Ueber den Begriff der Instinkhandlung," in *Folia Biotheor, Leiden* 2:17–50.

_____ (1937b). "Ueber die Bildung des Instinkbegriffes," in *Naturwissenshaften* 25:289–300, 307–18, 324–31.

_____ (1965). *Evolution and Modification of Behavior*. Chicago: University of Chicago Press.

McNew, S. (1984). "Hyperactivity: The Implications of Heterogeneity," in *Malformations of Development: Biological and Psychological Sources and Consequences,* ed. E. S. Gollin. New York: Academic Press.

Mead, George Herbert (1956). *George Herbert Mead on Social Psychology,* ed. Anselm Strauss. Chicago: University of Chicago Press.

Miller, John W. (1978). *The Paradox of Cause and Other Essays.* New York: Norton.

_____ (1980). *The Definition of the Thing, with Some Notes on Language.* New York: Norton.

_____ (1981). *The Philosophy of History, with Reflections and Aphorisms.* New York: Norton.

_____ (1982). *The Midworld of Symbols and Functioning Objects.* New York: Norton.

_____ (1983). *In Defense of the Psychological.* New York: Norton.

Naess, Arne (1973). "The Shallow and the Deep, Long Range Ecology Movement: A Summary," in *Inquiry* 16:95–100.

Nagel, Thomas (1979). *Mortal Questions.* Cambridge: Cambridge University Press.

_____ (1986). *The View from Nowhere.* New York: Oxford University Press.

Ortega y Gassett, José (1941). *Toward a Philosophy of History,* trans. Helene Wyle. New York: Norton.

Overton, Willis F., ed. (1983). *The Relationship Between Social and Cognitive Development.* Hillsdale, N.J.: Erlbaum.

Parsons, Talcott (1951). *The Social System.* Glencoe, Ill.: The Free Press.

Piaget, Jean (1952 [1936]). *The Origins of Intelligence in Children,* trans. M. Cook. New York: International University Press.

_____ (1965). *The Moral Judgement of the Child,* trans. Marjorie Gabain. New York: The Free Press.

_____ (1971). "The Theory of Stages in Cognitive Development," in *Measurement and Piaget,* ed. D. R. Green, M. P. Ford, and G. B. Flamer. New York: McGraw-Hill.

Pipp, Sandra, M. Ann Eastwood, and Scott R. Brown (1993). "Attachment Status and Complexity of Infants' Self- and Other-Knowledge When Tested with Mother and Father," in *Social Development* 2 (1):1–14.

Plato (1961). *The Complete Dialogues, Including the Letters,* ed. Huntington Cairns and Edith Hamilton. New York: Pantheon Books.

Reese, H. W., and W. F. Overton (1970). "Models of Development and Theories of Development," in *Life-Span Developmental Psychology: Research and Theory,* ed. L. R. Goulet and P. B. Baltes. New York: Academic Press.

Russell, Robert L. (1991). "Narrative in Views of Humanity, Science, and Action: Lessons for Cognitive Therapy," in *Journal of Cognitive Psychotherapy: An International Quarterly* 5 (4):240–56.

Sahlins, Marshall (1977). *The Use and Abuse of Biology.* Ann Arbor: University of Michigan Press.

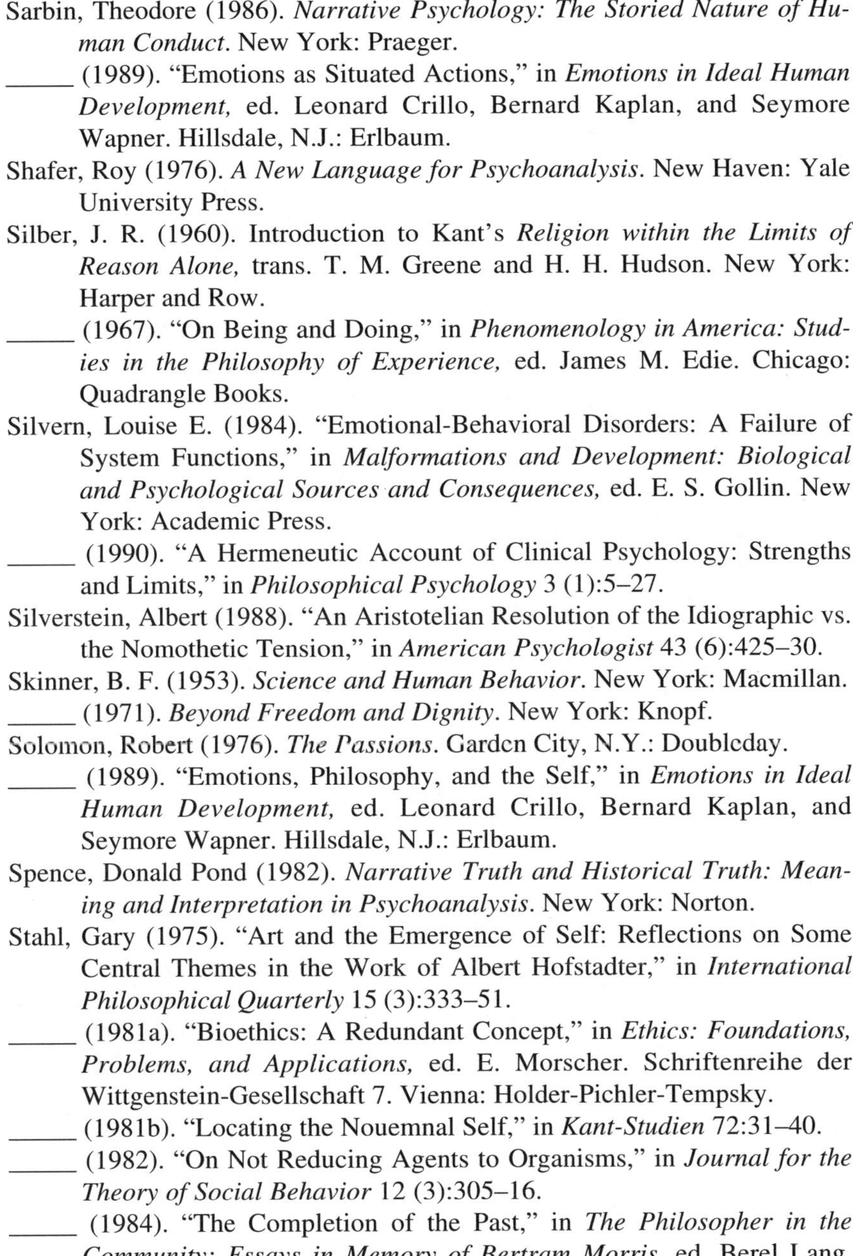

Sarbin, Theodore (1986). *Narrative Psychology: The Storied Nature of Human Conduct*. New York: Praeger.

_____ (1989). "Emotions as Situated Actions," in *Emotions in Ideal Human Development,* ed. Leonard Crillo, Bernard Kaplan, and Seymore Wapner. Hillsdale, N.J.: Erlbaum.

Shafer, Roy (1976). *A New Language for Psychoanalysis*. New Haven: Yale University Press.

Silber, J. R. (1960). Introduction to Kant's *Religion within the Limits of Reason Alone,* trans. T. M. Greene and H. H. Hudson. New York: Harper and Row.

_____ (1967). "On Being and Doing," in *Phenomenology in America: Studies in the Philosophy of Experience,* ed. James M. Edie. Chicago: Quadrangle Books.

Silvern, Louise E. (1984). "Emotional-Behavioral Disorders: A Failure of System Functions," in *Malformations and Development: Biological and Psychological Sources and Consequences,* ed. E. S. Gollin. New York: Academic Press.

_____ (1990). "A Hermeneutic Account of Clinical Psychology: Strengths and Limits," in *Philosophical Psychology* 3 (1):5–27.

Silverstein, Albert (1988). "An Aristotelian Resolution of the Idiographic vs. the Nomothetic Tension," in *American Psychologist* 43 (6):425–30.

Skinner, B. F. (1953). *Science and Human Behavior*. New York: Macmillan.

_____ (1971). *Beyond Freedom and Dignity*. New York: Knopf.

Solomon, Robert (1976). *The Passions*. Garden City, N.Y.: Doubleday.

_____ (1989). "Emotions, Philosophy, and the Self," in *Emotions in Ideal Human Development,* ed. Leonard Crillo, Bernard Kaplan, and Seymore Wapner. Hillsdale, N.J.: Erlbaum.

Spence, Donald Pond (1982). *Narrative Truth and Historical Truth: Meaning and Interpretation in Psychoanalysis*. New York: Norton.

Stahl, Gary (1975). "Art and the Emergence of Self: Reflections on Some Central Themes in the Work of Albert Hofstadter," in *International Philosophical Quarterly* 15 (3):333–51.

_____ (1981a). "Bioethics: A Redundant Concept," in *Ethics: Foundations, Problems, and Applications,* ed. E. Morscher. Schriftenreihe der Wittgenstein-Gesellschaft 7. Vienna: Holder-Pichler-Tempsky.

_____ (1981b). "Locating the Nouemnal Self," in *Kant-Studien* 72:31–40.

_____ (1982). "On Not Reducing Agents to Organisms," in *Journal for the Theory of Social Behavior* 12 (3):305–16.

_____ (1984). "The Completion of the Past," in *The Philosopher in the Community: Essays in Memory of Bertram Morris,* ed. Berel Lang,

William Sacksteder, and Gary Stahl. Lanham, Md.: University Press of America.

_____ (1986). "Remembering the Future," in *Nuclear Weapons and the Future of Humanity,* ed. Abner Cohen and Steven Lea. Totowa, N.J.: Rowman and Allanheld.

_____ (1990). "Making the Moral World," in *The Philosophy of John William Miller,* ed. Joseph Fell, pp. 111–22. *Bucknell Review* 34 (1). Cranbury, N.J.: Associated University Presses, Inc.

Strawson, P. F. (1959). *Individuals: An Essay in Descriptive Metaphysics.* London: Methuen.

_____ (1966). *The Bounds of Sense: An Essay on Kant's Critique of Pure Reason.* London: Methuen.

Stevenson, C. L. (1943). *Ethics and Language.* New Haven: Yale University Press.

_____ (1963). *Facts and Values: Studies in Ethical Analysis.* New Haven: Yale University Press.

Thomas, E. B., C. Acebo, and P. T. Becker (1983). "Infant Crying and Stability in the Mother-Infant Relationship: A Systems Analysis," in *Child Development* 54:653–59.

Thompson, Janna (1990). "A Refutation of Environmental Ethics," in *Environmental Ethics* 12:147–60.

Tillich, Paul (1957). *Dynamics of Faith.* New York: Harper and Row.

Tinbergen, Niko (1973). *The Animal and Its World.* Vol. 2. Cambridge: Harvard University Press.

Toulmin, Stephen (1970). "Reasons and Causes," in *Explanation in the Behavioral Sciences,* ed. Robert Borger and Frank Cioffi. Cambridge: Cambridge University Press.

_____ (1981). "Epistemology and Developmental Psychology," in *Developmental Plasticity: Behavioral and Biological Aspects of Variations in Development,* ed. E. S. Gollin. New York: Academic Press.

Urmson, J. O. (1988). *Aristotle's Ethics.* Oxford: Basil Blackwell.

von Uexkull, Jakob Johann (1926). *Theoretical Biology.* New York: Harcourt Brace.

_____ (1957). "A Stroll Through the Worlds of Animals and Men: A Picture Book of Invisible Worlds," in *Instinctive Behavior,* ed. C. H. Schiller. New York: International Universities Press.

Weiss, Paul (1971). "The Basic Concept of Hierarchic Systems," in *Hierarchically Organized Systems in Theory and Practice,* ed. P. A. Weiss. New York: Hafner.

_____ (1973). *The Science of Life.* Mt. Kisco, N.Y.: Futura.

Wilden, Anthony (1980). *System and Structure.* London: Tavistock.

Williams, Bernard (1981). *Moral Luck.* Cambridge: University of Cambridge Press.

_____ (1985). *Ethics and the Limits of Philosophy.* Boston: Harvard University Press.

Winnicott, D. W. (1953). "Transitional Objects and Transitional Phenomena," in *International Journal of Psycho-Analysis* 34:89–97.

Wittgenstein, Ludwig (1953). *Philosophical Investigations,* trans. G. E. M. Anscombe. Oxford: Basil Blackwell.

Wolff, Robert Paul (1969). *The Ideal of the University.* Boston: Beacon Press.

World Health Organization (1946). *Constitution.* Geneva: WHO.

Index